A Note on The Author

Tim Smedley is an award-winning sustainability journalist, based in the UK. At first, air pollution was just another environmental story to file. But air pollution got under his skin and into his veins – literally, as it turned out – and the more he researched, the more concerned he became.

Tim has written for the *Financial Times*, the *Guardian*, *The Sunday Times*, *New Scientist*, and the BBC. *Clearing the Air* is his first book.

Also available in the Bloomsbury Sigma series:

CLEARING
THE AIR

The Beginning and the
End of Air Pollution

Tim Smedley

BLOOMSBURY SIGMA
LONDON · OXFORD · NEW YORK · NEW DELHI · SYDNEY

BLOOMSBURY SIGMA
Bloomsbury Publishing Plc
50 Bedford Square, London, WC1B 3DP, UK

BLOOMSBURY, BLOOMSBURY SIGMA and the Bloomsbury Sigma logo
are trademarks of Bloomsbury Publishing Plc

First published in the United Kingdom in 2019
This edition published 2020

A catalogue record for this book is available from the British Library

Library of Congress Cataloguing-in-Publication data has been applied for

ISBN: PB: 978-1-4729-5333-9; eBook: 978-1-4729-5330-8

2 4 6 8 10 9 7 5 3 1

Typeset by Deanta Global Publishing Services, Chennai, India
Printed and bound in Great Britain by CPI Group (UK) Ltd, Croydon CR0 4YY

To find out more about our authors and books visit www.bloomsbury.com
and sign up for our newsletters

For Robin and Silvia,
Isaac and Thomas

Contents

Prologue

My first daughter was born in London in March 2014, at St Mary's Hospital in Paddington, on the first day of spring. The sky was a brilliant blue and daffodils burst yellow with optimism from grey concrete pavement pots. I staggered out into the mid-morning to a new day, a new life, with all the hopefulness of fatherhood and new beginnings (and an urgent order for coffee to fulfil). But I was oblivious to the fact that I was walking down one of the most polluted roads in one of Europe's most polluted cities. I also didn't know that we were in the middle of a month-long air pollution episode that would cause 600 deaths in London plus 1,570 emergency hospital admissions – some at the very same hospital I had just emerged from.

In the coming weeks and months of that year, however, I started to become aware of air pollution. I hate myself for fitting the cliché, but yes, parenthood did suddenly make me more aware of risk. The scene in the film *Paddington* where the Browns arrive at the maternity ward as carefree hippies on the back of a Harley-Davidson and leave nervously in a grey family car with their new arrival, rings embarrassingly true. As a sustainability journalist I had covered environmental issues for years for the *Guardian*, *Financial Times*, BBC and the *Sunday Times*. But I hadn't paid much attention to the most immediate environmental issue of all: the air we breathe. If I thought of air pollution at all, I thought of smog – and smog was something other countries suffered from. But now, with an infant to protect,* I started to notice the visceral throb of traffic in central London; the brown tinge to the sky even on cloudless days; the blackened nostrils after commuting on the Tube.

---

* And I *totally* agree that you don't need to be a parent to notice these things. In fact, more shame me for failing to act before passing on my genes.

My awakening seemed to happen in parallel with that of other Londoners, too. London's Mayor Boris Johnson confessed to the *Evening Standard* in December 2014 that Oxford Street had the worst diesel pollution on Earth. To say this came as a surprise is something of an understatement: the shopping street where I took my daughter to pick out her first pram had some of the most polluted air *in the world*?! Where were the health warnings, the public information signs, the protesters marching? All I could see were happy, oblivious shoppers. Barely a week into 2015 came another *Evening Standard*[*] headline: 'Oxford Street pollution levels breached EU annual limit just four days into 2015'. The article said things about 'the EU limit for nitrogen dioxide (NO_2) levels ... above 200 micrograms per cubic metre' and something regarding 'tackling diesel pollution'. But I had no idea what any of those things meant. What is nitrogen dioxide? Why is it bad? And what's the beef with diesel?

My professional instincts kicked in and I began to research. This started with a few articles, for the BBC and others, but soon the topic blew up. And not just in the UK, but in cities across the world. The smog in Beijing, China, was becoming so bad that it was dubbed the 'Airpocalypse'. Pictures circulated on social media of Beijing students sitting their exams so couched in smog that they could barely see the neighbouring table. Stories emerged from Delhi, India, such as the *Guardian*'s 'Toxic smog covers Delhi after Diwali' (31 October, 2016), which pointed the finger at 'the density of some harmful particles and droplets in the air ... up to 42 times the safe limit'. Particles of what, I wondered? Droplets of what? What safe limit? And then came the death figures. In late 2016, the World Health Organisation (WHO) announced that outdoor air pollution caused over 3 million deaths worldwide; by 2018, the WHO revised this up to 4.2 million.[†] In the UK, according

---

[*] I don't have any affinity with the *Standard,* but as a Londoner, with vendors at every tube station, its headlines are unavoidable.

[†] 7 million, if you include indoor air pollution. This book will largely focus on outdoor air pollution, however, because it is the one thing we all have in common, and that we have a collective responsibility to fix.

to the Royal College of Physicians, 40,000 people were dying each year from air-pollution-related illnesses. But what were these airborne illnesses?

I decided I had to go to these cities and meet the experts who could tell me what the hell was going on. And when I did, the depth and breadth of what they had to say made it clear that this was not going to be a series of articles – this was a book. In the words of the eminent American epidemiologist I spoke to, Dr Devra Davis, 'those felled by environmental conditions seldom even know why they are dying'. It's about time they were told.

Some of the stories I encountered were painful to hear. Beijing citizens told me of smogs so bad that the sky turned black in the middle of the day, and persistent illnesses that never went away. When I landed in Delhi in late 2017, I experienced some of the worst smog I'd ever seen or inhaled – and yet the general mood in the city was jubilant, because the conditions were far better than the week before, when even the street dogs started dying from heavy smoke that clung to the ground and entered people's homes like an intruder. The Delhi half marathon went ahead that week, regardless, with runners wearing face masks and complaining of burning eyes.

Mexico City was more polluted than Delhi in the 1990s. Melba Pria, now the Mexican ambassador to India, told me of her experiences as a young adult in the city back then, when the crisis peaked: 'The Ministry of Education tested sixth-graders [11-year-olds], and 80 per cent of the children in Mexico City said that the sky was grey and 10 per cent said the sky was brown. Only 10 per cent said that the sky was blue ... Then birds started falling from the sky. Little birds. Suddenly you found dead sparrows in the walkway ... In Mexico City despite everything else, you have lots of hummingbirds. Then there were no hummingbirds to be seen. It was like, "These are our birds, how is it that they are not there any more?!"' The smog was so bad that all the schools were closed for two consecutive months. 'I remember a colleague of mine, and I remember it very clearly because it was very shocking for me,' says Pria.

'Her kid was at home [during the school closure]. She was telling me that her kid was locked in the apartment. The boy said one day, "Mummy, my window friend is ill today". The mum said, what do you mean, "window friend"? The boy had made friends with another boy from another building that was relatively far away – they would play together by making signs from window to window, because they weren't allowed to leave the house. That was very disturbing to me.'

Yet I also found a powerful message of hope. Today, Mexico City has cleaned up its act. It ranks way down the WHO's list of most polluted cities – down in the 900s in fact, with air comparable to the medieval Italian city of Carpi and the famous cycling destination of Roubaix in France. Cities have reached untenable smog levels before and pulled themselves back from the brink. Through political and public will, they found effective solutions. In the first half of the twentieth century, the burning of dirty coal in domestic fires and inner-city power stations conspired to shorten the lives of almost all urban residents, culminating in the US in the Donora, Pennsylvania, disaster of 1948,* and London's Great Smog of 1952. The outcry, legislation and behaviour change that followed led to the UK Clean Air Act of 1956, a global landmark in environmental legislation, and the US Clean Air Act of 1970.

Unlike in Donora or the Great Smog, however, today's smog is largely invisible – the thick coal smoke replaced by tiny particles and chemicals. Modern science is starting to reveal what our eyes cannot see: an anonymous killer born from the cars in our driveways and the industrial processes used to make the products in our cupboards. But most people don't read modern science journals. Parents on the school run in their SUVs have never been told that the pollution inside their car can be four to five times worse than that on the street outside. Or that decades of studies in Europe and America show that air pollution stunts lung growth in

---

* I look at the Donora disaster in more detail in Chapter 4.

children. Or that the air pollution affects us at every stage of our life, from reducing our fertility levels to causing heart attacks and dementia.

Nick, a London cyclist hospitalised due to long-term exposure to air pollution, told me: 'Pollution is a slow, grinding, background thing, and people aren't very good at reacting to slow, grinding, background things. We all have our heads in the sand about this. And it's made worse by the fact that we can't see it. It needs to be made real to people.' We still have some visible plumes of coal smoke, notably in Poland, India and China. But the common foe that unites all cities is cars. I discovered that even the newest cars give off large exhaust plumes of invisible nanoparticles and nitrogen dioxide gas – that are amongst the deadliest adversaries we face. David Newby, a professor of heart medicine, tells me in Chapter 3 how nanoparticles appear within our bloodstream, clogging up our arteries and causing high blood pressure and heart attacks. These traffic-derived pollutants appear in each and every street with cars. They will appear in your street.

A refrain I came across a lot is the 'it never did us any harm' argument. That 'when we were kids we drove X or burnt Y and look, we're fine'. But often, if you dig a little deeper they admit to some lingering health problem such as asthma, persistent lack of fitness, hay fever, high blood pressure or worse. For example, when I was a child in the 1980s every car on the road ran on petroleum mixed with lead. Childhood lead exposure is known to reduce the parts of the brain responsible for mood regulation and decision-making, leading to reduced impulse control and more aggressive behaviour. The natural background level of lead in human blood is around 0.016 micrograms of lead per decilitre of blood ($\mu g/dl$). Due to the global use of leaded petrol however, from 1976 to 1980 the median blood lead level of US children aged under five was $15\mu g/dl$ – almost a thousand times higher than it should be. From 1979, researchers in Cincinnati recruited pregnant women, took blood tests from babies and then repeated them every year until the children were six and a half years old. When these data were overlaid with criminal records up to October 2005 (this was a very long study), they found that increased blood lead levels were associated with higher rates of arrest, particularly for violent crimes – unsurprising, really, if you're poisoning them with something that inhibits mood control. For every 5 µg/dl increase in blood lead levels at six years of age, the risk of being arrested for a violent crime as a young adult in the US increased by almost 50 per cent.[1]

Similar 'lead and crime' studies have since been carried out in other countries including the UK and Australia, all broadly coming to the same conclusion: even after adjusting for major demographic variables (age, education, income, etc.), lead in the air remained the largest determinant of variance in violent criminality. Uta Frith, Emerita Professor of Cognitive Development at University College London, studied the impact of lead poisoning in children living near busy roads in east London in the 1980s. She recalled for the BBC in 2018 that, 'the results were overwhelming and could not be dismissed as due to other factors such as parent's socio-economic status … there was a detrimental effect on children's cognitive abilities and behaviour, and decreases in IQ test scores could be precisely linked to increases in the amount of lead in the blood … That was a revelation'.[2]

The story of lead pollution is a modern parable. It begins with a single greedy corporation looking to make profit from a known poison (I'll tell that story in Chapter 3). It also shows the power and reach of a single type of pollution from a single source: the automobile engine. Owing to that one minuscule period in the Earth's history, a thin layer of lead now covers the entire planet. Some of which came from my parents' 1980s Volvo. But like Mexico City, leaded petrol also has a happy ending. It is one of the few examples of the international community coming together, understanding a major health issue and banning the pollution source that caused it. Most developed countries phased out leaded petrol, specifically because of its detrimental health effects, in the 1980s and

1990s. The levels of lead in children's blood in the US decreased by 84 per cent from 1988 to 2004. By 1999 the median blood level among US kids was down from 15µg/dl to just 1.9µg/dl.*

Other common refrains I came across are that 'there is no one-size-fits-all solution', and 'of course *this* city is different'. But in fact, as my research progressed, a blueprint did emerge for a clean-air city. I found that my outsider 'non-expert' status quickly became a strength: academics and policy makers from other areas talk to each other a lot less than you might think, pushed into silos and specialisms by the nature of their work. As someone memorably told me during my research, 'When I go to conferences, all the epidemiologists sit on one side of the room, and all the toxicologists sit on the other side.' By my visiting both sides of the room, across many different disciplines and countries, an action plan formed that others can follow. Throughout the pages of this book, I'm going to spell out what it is. But here's a spoiler: if you think (as I used to) that the stuff you burn, from a cosy log fire to the fuel in an engine, isn't doing you or your neighbours any harm, then you're in for a shock.

I describe the chemical cocktail of pollutants in our air in Chapter 2, but there is one I need to introduce from the very start, because it's going to crop up a lot: particulate matter (PM). These are the tiny solid particles that float in the air, from road dust to smoke, that do us the most long-term damage. Scientists define PM not by what the particles are made of (i.e. coal smoke, agricultural dust, engine fumes) but by size. The large ones are called PM10 – so named because it refers to any particle measuring 10 micrometres in diameter or below, roughly a tenth of the width of a human hair. PM10s can be seen by the naked eye as smoke or haze. In typical conditions, PM10s are also easily filtered out by our

---

* Albeit still a hundred times greater than the natural level. According to toxicologists, there is no known 'safe level' of lead pollution.

body's natural defences, such as nose hair. Smaller PM2.5s, however, with each particle less than 2.5 micrometres in diameter, are a different matter, and their little sisters the 'nanoparticles' even more so. Typically the result of modern combustion techniques, they are much too tiny to see even in large quantities, can bypass our bodies' defences, and pass through the walls of our lungs and directly into the bloodstream. It's not a spoiler to say that PM2.5 will be a major focus throughout the book, as indeed it is a major focus of modern air pollution science and regulation.

Comparing the PM2.5 levels in different cities is not always straightforward. PM is measured by the mass of particles found per cubic metre of air ($\mu g/m^3$), and while the WHO recommends a health-based limit of $25\mu g/m^3$, most countries and regions have their own limits, such as $50\mu g/m^3$ in the EU or $65\mu g/m^3$ in the US. The official monitoring stations used to capture these measurements are often too few in number, or poorly positioned – in many cities I came across a public suspicion that they are purposely placed in relatively clean air locations, in order to skew the numbers. So, while I do refer to official readings where available, on my travels for this book I also carried a portable PM2.5 monitor called the 'Laser Egg 2'. I could turn my 'Egg' on and check the PM2.5 $\mu g/m^3$ reading wherever I went and, while levels in one street can be very different from levels in another, it was a useful reference tool.

And there's a gas I want to deal with from the start, too. Or rather, a category of gases: greenhouse gases. Local, ambient* air pollution differs from climate change in a number of important ways. We can breathe in the main greenhouse gases, carbon dioxide (CO_2) and methane (CH_4), in surprisingly high quantities in the air with no real harm to our health. Equally, the common air pollutants that cause us

---

* Ambient is just another word for 'outdoor' air, as distinct from indoor, and is used by the likes of the WHO. It needs mentioning, as many voluminous reports only ever use 'ambient', but I'll mostly just say 'outdoor' from now on.

the biggest health problems – namely PM2.5, surface-level ozone and nitrogen dioxide – tend to have negligible global warming effects (although there are exceptions, as we shall discover). The easiest way to separate the two issues, however, is also what attracted me to writing about air pollution: it is a local problem, and solvable locally. While carbon emissions from one country contribute to the changing climate globally, the same is not so true of air quality. There are some 'transboundary' air quality issues, with one country's pollution blowing over their borders into neighbouring states, but for the most part the pollution is hyper-local. The smallest particles, nanoparticles, only exist within metres of their source; the lifespan of nitrogen dioxide is typically no more than a day, and often much less, meaning it can't get very far – you won't find any in remote rural regions; surface-level ozone is so highly reactive it can disappear within hours. So, if your town or city carries out all the measures outlined in my blueprint, you *will* breathe in cleaner air, irrespective of what your neighbouring state does, or what countries on the other side of the world get up to. Even if you only convince the people on your street to follow the blueprint, while the rest of your town continues to pollute, the air on your street will be measurably cleaner than other, neighbouring streets. And it just so happens that most of the measures to reduce outdoor air pollution tend to reduce greenhouse gas emissions and tackle climate change too.

When I walked out of that London hospital in 2014, however, I didn't know any of that. As I nervously helped my recovering wife and our one-day-old daughter towards the awaiting grey family car, we were exposed to a high air pollution episode that very day. The pollution episode of March 2014 wasn't actually reported accurately until December 2016, in the specialist science journal *Environment International*. The researchers revealed that a pollution peak had been dismissed by the government at the time as merely a cloud of 'Saharan dust' – a natural phenomenon and an easy scapegoat. However, according to the journal authors it was in fact nitrogen dioxide caused by London traffic pollution

mixed with agricultural ammonia from nearby farms –
Saharan dust made up less than 20 per cent of it. When I read
that, I was both angry (how could a government mislead its
people on such an important issue?) and intrigued (how can
cow pee from farms foul up the air?). I had a lot to learn. It
proved to be an exciting, at times blood-boilingly infuriating,
but ultimately hope-filled journey.

PART ONE
ORIGINS

CHAPTER ONE

The Greatest Smog?

London: 1952

Both my parents were born in 1952. My maternal grandfather, Selwyn Bate, was a local solicitor in Tamworth, my home town, and a lecturer at Birmingham Law School. On occasion – and despite his best efforts to avoid it – he had to travel into London. My aunt Claire remembers him once having to take the train to London carrying an Elizabethan legal charter, written on sheep's hide, to settle a dispute in the High Court. He hated the capital. Too many people and too dirty. He would wash as soon as he returned home. I used to assume when hearing these stories that it was down to a kind of inverse snobbery. As a Lancastrian, he perhaps considered Tamworth – in the Midlands – to be as far south as he wanted to go. Then I learned of the Great Smog of 1952, and suddenly, the fact that Grandpa Bate avoided London like the plague made a lot of sense.

The early winter weather in London in 1952 had been very cold, with heavy snowfall spreading across the whole of southern England. To keep warm, Londoners would burn large quantities of coal in their homes. During a particularly cold spell like this one, most households would 'bank up' the fire at night with coal dust so that it would last until morning. Electricity was provided by inner-city power stations at Battersea and the South Bank, burning coal and belching smoke from chimneys taller than cathedral spires. In winter, not only were people burning more fuel and creating more smoke, but conditions for a 'temperature inversion' would arise: if the air close to the ground is cooler than the air higher above, it becomes trapped. With a clear, windless sky and moist, damp ground, a fog also forms. And London had long been famous for them: 'pea-soupers' in Cockney slang, romanticised in the stories of Dickens and Conan Doyle and paintings by Turner and Monet (who preferred to visit in winter, transfixed by the smog's swirling yellow light).

On Friday 5 December 1952, the familiar thick fog appeared once again across London. But this one didn't lift the following day, or the next. A prolonged temperature inversion saw smoke concentrations reach 56 times the 'normal' level. The official records show visibility reduced in some places to just one yard (91cm), the lowest ever recorded in the city. People couldn't even see their own feet. Blinded commuters stepped off bridges into the icy Thames and from railway platforms into the path of oncoming trains. Within 12 hours of the smog's arrival, people in their thousands began having respiratory problems and hospital admissions increased dramatically.

According to Met Office records, the smog spread for 13 miles and contained the following pollutants: 1,000 tonnes of smoke particles (which we'd now call 'black carbon' or PM10), 2,000 tonnes of carbon dioxide, 140 tonnes of hydrochloric acid and 14 tonnes of fluorine. Most hellish of all, 370 tonnes of sulphur dioxide were converted, suspended by the water droplets in the fog, into 800 tonnes of sulphuric acid. It hung like this, thick in the air, for five days.

At Sadler's Wells Theatre, a performance of *La Traviata* was stopped when the audience could no longer see the stage. At Smithfield livestock market, farmers wrapped whisky-soaked hessian masks over their cows to try and protect them. The smog stuck to windscreens like paint, forcing drivers to abandon their vehicles. The Fleet Street correspondent of the *Northern Whig* newspaper reported on Saturday 6 December that 'it has seeped into the shops and offices, where lights have been on all day … The smoky "peasoupers" not only damage the health of the inhabitants, but the insidious chemicals they contain eat into the stones and brick of buildings, and burn and blacken the trees.' There was also a nervous jokiness to the early reports. The *Daily Telegraph* reported the first casualty to be 'A mallard, presumably blinded by the fog, which crashed into Mr John Maclean as he was walking home in Ifield Road, Fulham. Both were slightly injured.' *The Times* leader column was even more dismissive, proclaiming that 'fogs are ancient Britons. They met the boat when the ancestors of Boadicea landed ... they roam about ... as freely as they did before anyone had heard of smoke abatement.' Perhaps taking their lead, the political response from Churchill's Conservative government was initially high-handed. Harold Macmillan, the Housing Minister, told the House of Commons that 'broad economic considerations needed to be taken into account' – in other words, the needs of industry come before worries about the weather.

Roy Parker, a third-year student at the London School of Economics at the time, remembered in 'The Big Smoke: Fifty years after the 1952 London Smog', a witness seminar held at the Centre for History in Public Health on 10 December 2002, that the early news reports were mostly concerned with the cancellation of sporting fixtures: 'neither I nor they paid much attention to the health consequences for people. I should have realised the full magnitude of the catastrophe, because my father, who was a driver on a steam locomotive, and had been partially gassed during the 1914–18 war, had many of the symptoms one associates with ... the inhalation of the coal dust and the sulphur. Most of the time he had difficulty breathing. He was 56 years old, and when I saw him that weekend he was in great distress, gasping for breath, struggling, but insisting that he would go to work on his bicycle ... It's hard now to appreciate how general the experience of chronic bronchitis was for the industrial working class in this country ... in my family, all the men had symptoms of this kind.'[1]

By the fourth day, the tone had changed. The *Hartlepool Northern Daily Mail* called it the 'great black-out fog ... so dense that most police patrol cars were immobilised, and police answered 999 calls by going on foot.' The cattle at Smithfield market were now dying, with others being prematurely slaughtered 'at the request of the owners'. A Fleet Street reporter for one of the Monday morning papers declared: 'This is no mere pea soup: everything is in it – the hors d'oeuvre, the fish, the joint, the sweet, the savoury, the black coffee and the waiter's scowl. It stings the eyes, and fouls the breathing … In Trafalgar Square I could hear the fountains, but they were invisible … [a colleague] found a man searching for a Tube station in the middle of Blackfriars Bridge.'

When the fog lifted on the fifth day, Tuesday 9 December, reports of the overflowing hospitals started to filter through, and the significance of the event began to dawn. As at the Donora, Pennsylvania disaster, just four years previously, funeral parlours ran out of coffins and florists of the flowers to accompany them. Due to the traffic congestion and near-zero visibility, many more died at home than had made it to hospital. Rosemary Merrit recalled for the *BBC World Service* in 2012 that her father walked a mile and a half home from work through the smog. That night he 'coughed so hard that he turned blue, and my mother woke up the neighbours to help … we couldn't get him to hospital because there were no ambulances.' He died the next day. His body was kept in the front room of their house for three weeks until an overworked undertaker was finally able to bury him, just before Christmas. 'I never liked to be in the front room after that,' she said. 'It was always very cold.'

In that one week alone, 4,703 people died in London – 3,000 more than usual. An environmental catastrophe, the city's own, famous fog, had caused more civilian casualties in London than any five-day German bombing campaign had managed just a few years before. It wasn't only the elderly and infirm who were struck down. One ambulance carried a 21-year-old sailor on active service. The doctor treating him, Dr Horace Pile, later recalled for the UK Channel 4 documentary *Killer Fog* in 1999 that he had 'never seen anything like it from a young man of that age, with breathing difficulties and a desperately failing heart'. When the ambulance reached the hospital it was already full up with victims. So was the second. When the ambulance set off in search of a third, the sailor died en route. The eventual number of fatalities is believed to have reached some 8,000 to

12,000, plus thousands more cases of lifelong health problems including lung and heart diseases.

In the 1930s and 1940s, London's above-ground public transport was largely served by emission-free electric trams, including double-deckers. However, they were rapidly usurped by the internal combustion engine. The last London electric tram was replaced by a diesel bus on 5 July 1952, just five months before the Great Smog. By December, 8,000 new diesel buses were on the roads, adding their fumes to the winter fog. Dr Barry Gray, a chest physician at Kings College Hospital who was interviewed for the *Killer Fog* documentary, described the change from electric trams to diesel buses in the 1950s as 'a disaster that has had a huge impact on the health of Londoners'.

After Roy Parker's personal experiences as a student during the Great Smog, he dedicated his career to air pollution research. He calculated that there were about 12 million domestic coal fires in the UK in 1952. There were also 20,000 steam trains burning low-quality coal, while Battersea power station alone burnt 10,000 tonnes of coal every week. The idea of having no petrol or diesel cars on our roads today seems as hard to imagine as the banishment of steam trains and coal fires did to a Londoner in the 1950s. But remarkably, in little more than a decade, that's exactly what happened. In 1953 a Committee on Air Pollution reported that there was a clear link between pollution and respiratory diseases, laying the foundations for the UK's Clean Air Act (1956). A complete modernisation of Britain's railway network was announced in 1955, effectively bringing to an end the era of the coal-fired steam engine. Battersea power station eventually closed in 1975. By the late 1970s, London's 'pea-soupers' were a thing of the past. Professor Peter Brimblecombe chaired the witness seminar at the Centre for History in Public Health in 2002. I asked him if there was a general feeling at that event, on the 50th anniversary of the Great Smog, that this problem had now been solved. 'Oh definitely,' he replied immediately. 'I think there was definitely that feeling … they believed it had been transformed enormously.'

LONDON: 2010s

On a cold April morning in 2016, in London's Trafalgar Square, a protester climbed up the city's iconic landmark, Nelson's Column. Alison Garrigan had been planning her ascent for months. In the watery light of sunrise, she and an accomplice fitted a gleaming white gas mask to Admiral Lord Nelson's blackened, sooty face. As a veteran Greenpeace activist, she hoped to highlight the state of London's air pollution. The Royal College of Physicians and the Royal College of Paediatrics and Child Health had recently estimated that up to 40,000 premature deaths were caused by air pollution in the UK each year, 10,000 of whom were Londoners. Children in some polluted parts of London had 5 to 8 per cent lower lung capacity than average in their age range. In 2013, nine-year-old Ella Kissi-Debrah, from south London, died following repeated hospital admissions for asthma, leading to calls for an inquest into the part air pollution played in her death[*]. In January 2017, Baroness Jones of the House of Lords levied the accusation of 'criminal neglect by a government that has no desire to protect the health of its citizens' in an article in *The Times* headlined 'Calls for curbs on traffic after 10 days of smog left 300 dead'.

So, given that London had supposedly solved this problem in the 1970s, what had gone so badly wrong? In a 1972 conservation handbook 'The New Battle Of Britain', the author H. F. Wallis fires a warning flare from history: 'Los Angeles-type "smog" – caused by the action of sunlight on petrol fumes – is possible in Britain ... new cars on our roads will still be able to poison the air to the same extent as now and, of course, there will be many more of them.'[2] Wallis

---

[*] In July 2018 Stephen Holgate, professor of immunopharmacology at the University of Southampton, submitted evidence in support of an inquest, stating a 'real prospect that without illegal levels of air pollution Ella would not have died' and a 'firm view' that her death certificate should include air pollution as a causative factor.

noted that between 1957 and 1967 'the British Rail passenger network slumped from 14,622 miles to just under 10,000 miles' while one-fifth of bus services disappeared. People had no choice but to switch to private cars. At his time of writing, there were just 14 million cars on British roads. By the end of 2017, there were 37.7 million licensed vehicles in Great Britain, around 31 million of which were cars. And 12.4 million of those – almost as many as all the cars in Wallis's day – were diesel cars.

When Professor David Newby began studying air pollution in the early 2000s, he assumed – like the Great Smog survivors – that the health impact of air pollution was yesterday's news. By the 1990s and 2000s, the Victorian inner-city coal power stations that once choked London had metamorphosed into cultural centres; the Tate Modern and Battersea power station now pumped out tourist dollars, not smoke. They stood as towering testaments to our more enlightened times. But Newby, now a professor of cardiology at the British Heart Foundation Centre of Research Excellence, was about to undertake a study that would completely transform his, and our, understanding of air pollution. In 2007, Newby and his team took healthy volunteers, put them in an exposure chamber, strapped them to a cycling machine and told them to start pedalling. The chamber was then filled with diesel exhaust fumes. 'People did question my ethics,' he admits. 'But I pointed out that the pollution levels outside on the street they walked down to get to the study were probably worse.' What they discovered was truly shocking. On exposure to ordinary street levels of vehicular exhaust fumes, blood became thicker and more likely to clot. Signs of stress on the heart were immediately apparent. Blood pressure went up and arteries visibly thinned. It was remarkably similar to cigarette smoking – something Newby calls 'self-induced air pollution' – except that, with street pollution, there is no option to quit. 'People were dying not of asthma but of heart disease and strokes caused by air pollution … Many people were surprised, including me.'

Poor air was not confined to the history books. In fact, there are more pollution particles in the air now than during London's Great Smog. The difference is that the particulate matter (PM) from modern pollution sources is too small to see.

Almost two-thirds of the population in the UK are living in areas with air pollution levels above EU legal limits. Greenwich councillor Dan Thorpe, a local primary school teacher by day and an elected Greenwich borough councillor the rest of the time, told me, 'My school, Windrush Primary School, is right by the Thames Barrier, so we've got the Woolwich Ferry and the roundabout at one end, the industrial estate behind ... they stack the heavy goods vehicles onto the ferry, and when they enter the roundabout all it takes is one minor incident and then the traffic backs up and you have a major pollution incident ... Once you get that it's a sort of Armageddon situation.'

Another long-time south Londoner, Nick Hussey, grew up cycle racing. While his friends were into football, his idols were the Tour de France winners Miguel Induráin and Greg LeMond. Since then he'd always been a cyclist, his racing dreams gradually giving way to the more ordinary life of a cycling commuter. In May 2005, aged 32 and living in London, his 'hay fever' started. 'I can remember the prickling, stinging eyes, not being able to breathe properly,' he tells me. 'Not long after that I started to develop problems with eating certain foods. I seemed to be having a big histamine battle. I began to find cycling a struggle, particularly in warmer months. But it didn't quite seem to match the hay fever pattern. I didn't really understand it.' Living close to a major road in south-west London, his breathing became so bad he twice ended up in A&E. 'The NHS actually pushed me through the system quite quickly to find out what was wrong. They found I actually had far larger than average lung capacity, so I had no right to have breathing problems. They took tests and narrowed down the cause. They ruled out pollen. They ruled out diet.' Finally, a specialist in a drab hospital consulting room delivered the news. 'He just said "Yep, it's air pollution: welcome to London." Like it was the most normal thing in the world, as if he saw it every single day.' He sounds upset when he tells me this. 'If it's that simple and obvious, why don't we know more about it? And why aren't people angrier about it?'

Under EU law, a city is allowed to breach hourly maximum limits for nitrogen dioxide (NO_2) up to 18 times in a calendar year. In 2016 it took just seven days for London to breach that annual limit. In 2017, it was down to five days, using up its annual allowance on 5 January. In October 2017, the government's plan to cut NO_2 pollution was so lacklustre that it was deemed illegal and inadequate by the High Court. Even the United Nations Human Rights Council got involved, producing a 22-page report on air pollution in the UK, which estimated that the impact of poor air quality cost the country about £18.6 billion per year in health costs. Its report in 2017 was damning: 'Air pollution continues to plague the United Kingdom ... Children, older persons and people with pre-existing health conditions are at grave risk of mortality, morbidity and disability, with magnified risks among the poor and minorities.'[3]

Rather than learning from the lessons of the Great Smog, in the 2000s London became the global epicentre for diesel exhaust fumes, one of the most potent sources of NO_2 and particulate matter pollution.* In 1950, there were around 35 million cars in the entire world. Today, there are nearly as many in the UK alone. Meanwhile domestic wood and coal fires, banished from inner cities by the Clean Air Act, are making an unlikely comeback. Re-branded as 'biomass burners' using 'renewable fuel', they have been actively encouraged. A government consultation document in 2018

---

* We'll learn more about both these pollutants in Chapters 2 and 3, while Chapter 5 looks specifically at diesel.

acknowledged that an 'increase in burning solid fuels in our homes is having an impact on our air quality and now makes up the single largest contributor to our national PM emissions'. In central London, wood burning accounts for up to 31 per cent of the urban-derived PM2.5 – probably for the first time since the 1950s.

The UN Human Rights Council report concluded, 'by neither taking action as expeditiously and effectively as possible, nor taking all possible measures to reduce infant mortality and to increase life expectancy, the United Kingdom Government has violated its obligations to protect life, health and the development of children in its jurisdiction.' Out of 51 UK towns and cities listed in the 2018 WHO air quality database, 44 breach the WHO recommended PM2.5 limits.*

The Beijing 'Airpocalypse'

In 2008, the US embassy in Beijing made a controversial decision. In a country where no official air pollution data was provided to its citizens, the US embassy installed an air pollution sensor – a MetOne BAM 1020 and Ecotech EC9810 monitor, to be precise – on its rooftop. Later that year, it set up a Twitter account, @beijingair, to automatically tweet the PM2.5 level on the hour, every hour. Despite repeated requests by the Chinese authorities to take it down, the embassy continued releasing its PM2.5 readings. Vice Minister of Environmental Protection Wu Xiaoqing told a press conference, 'Diplomats are obligated to respect and abide by the local laws and regulations ... We wish those embassies and consulates will respect China's laws and stop publishing air quality data.' Mark Toner, of the US State Department, retorted: 'We provide the American community,

---

* WHO health-based recommendations call for countries to reduce their air pollution levels to annual mean values of $10\mu g/m^3$ for PM2.5 and $20\mu g/m^3$ for PM10.

both our embassy and consulate personnel ... information it can use to make better daily decisions regarding the safety of outdoor activities.'*

The reasons for doing so were obvious. The skies across China, and the northern Hebei Province in particular, were getting greyer by the year. In 2004 the Chinese investigative journalist Chai Jing asked a small child whether they had ever seen stars? Her answer was no. Have you ever seen a blue sky? 'Slightly blue,' the child said. Have you ever seen white clouds? 'No.' By 2009, China's Environmental Protection Department carried out a pilot project in its major cities monitoring for 'haze' – a euphemism for smog, before smog was officially acknowledged – and discovered the number of annual haze days ranged from a low of 51 days in the year to a high of 211 days.

'When I first noticed it, when you could really start to see it, was in the late 1990s,' says American businessman Manny Menendez, when we meet in Beijing in December 2017. Manny has been working in China ever since it opened its borders to international trade, brokering the very first joint-venture deal between the US and China in 1980. 'I remember in the early days ... industry was in the heart of the city, because it was easy to access. From a sustainability or urban design point of view, that just doesn't work.' Decades of working in China, Manny admits, have caused him health problems. 'I get a wheeze. I have a mask and I will wear a mask. I don't wear it as much as I should. If the emissions are bad and it is really a high PM2.5 count ...' he pauses to cough when saying this, as if triggered by the thought, '... then I will get a wheeze that will last at least a week to 10 days. It won't go away.' He has business

---

* The US Department of State has since rolled the scheme out to over 20 US embassies worldwide, including the embassy in Delhi. On my travels, I typically found that the US embassy reading was the one most locals turned to as the most trusted source. You can find the live readings at www.airnow.gov.

associates who have moved their families out of Beijing because they 'didn't want them to grow up with respiratory problems'.

When the Institute of Public & Environmental Affairs (IPE), a Beijing-based NGO, published its first Air Quality Transparency Index (AQTI) in 2010, there were no daily air quality readings available for any city in China. The IPE report had to make use of industrial data and the scant figures included in the government's annual 'Report on the State of the Environment in China', which gave no granularity of detail beyond the percentage of (unnamed) cities that met the (undefined) national standards of grades 1 to 3. In response, the first IPE report didn't pull any punches, to the surprise of many in China and observers in the West: 'In recent years, air pollution has become one of the most pressing environmental problems faced by Chinese cities,' stated the IPE in 2010, becoming one of the first organisations to say so publicly. 'Bad air quality not only impacts the lives of hundreds of millions of urban residents but also threatens their health and safety. China is undergoing a period of rapid industrial and urban development, which is largely the cause of air pollution in many of its cities ... Currently, China has not yet formulated a comprehensive national air pollution and health monitoring network.' Unlike the vague government report, the IPE listed the main culprits: 'Coal smoke pollution ... Sulfur dioxide (SO_2) and total suspended particle pollution problems in urban areas ... the number of motor vehicles increasing the severity of exhaust pollution. Air pollution problems such as haze, photochemical smog and acid rain ... becoming more prominent day by day.'

Meanwhile the US embassy air quality monitor kept on tweeting. Twitter itself had been blocked by China's Great Firewall in 2009, but several mobile phone apps and Weibo (the Chinese version of Twitter) posted the @beijingair readings anyway. If PM2.5 readings went above 200 micrograms per cubic metre of air ($\mu g/m^3$), the @beijingair account would automatically issue a 'very unhealthy' warning; above $400\mu g/m^3$, it became 'hazardous'. Above $500\mu g/m^3$, however, was deemed so unlikely that it was carelessly programmed to announce that the air was 'crazy bad'. On 18 November 2010, the unlikely happened. The first 'crazy bad' warning was issued at 8 p.m., as the reading tipped into $503\mu g/m^3$ (it peaked the next day at $569\mu g/m^3$) – much to the embarrassment of both the US embassy and the Chinese authorities. The message was swiftly sanitised to 'beyond reading', but the damage was done. The 'crazy bad' message spread like wildfire. Chinese social media started to pick up on this and started re-tweeting this on Weibo,' recalls Kate Logan, director at IPE, when I visit IPE's Beijing office. 'Chinese citizens were starting to pay attention.'

Then came the 'Airpocalypse' of 2013. Whereas the US embassy had set the Twittersphere buzzing with a $500\mu g/m^3$ reading in 2010, on 12 January 2013 it was reporting readings of over $800\mu g/m^3$. Even the Beijing Municipal Environmental Protection Bureau, which had by now begrudgingly started to release some official readings, showed PM2.5 levels exceeding $700\mu g/m^3$. Liam Bates, a Swiss-British expat living in Beijing, recalls: 'It felt like 6 p.m. when the sun is setting and everything starts to go dark – but it was actually lunchtime. It was just nuts. It actually did feel like the apocalypse – it was the middle of the day, and the sky was going black ... people started really freaking out ... Even people who were previously sceptical were like, "OK, this is really messed up".'

From 10 to 14 January, the $\mu g/m^3$ reading never dropped below three figures; the whole month averaged around $200\mu g/m^3$. To put that into context, that is thicker than the air found inside the average airport smoking lounge ($167\mu g/m^3$, according to a 2012 US study). The WHO sets a health-based daily limit of just $25\mu g/m^3$. During the Airpocalypse, the *South China Morning Post* reported that Beijing Children's Hospital was receiving more than 7,000 patients a day, with the number of children being treated for respiratory ailments hitting a five-year high.

In the year of the Airpocalypse, the International School of Beijing, a private school for the children of expats and wealthy locals, erected a huge pressurised dome over its outdoor playground to keep out the city's now unbreathable air. The dome's structure is kept aloft by huge pressurisation fans that filter the air, allowing its fee-paying pupils to play 'outside' once more. Also that year, drink cans filled with compressed PM-free air, available in multiple 'flavours' including 'Pristine Tibet', were sold on the streets in Beijing. It would have been the perfect piece of performance art, were it not for the fact that the guy who made the cans, Chen Guangbiao, was an entrepreneur at the height of China's economic boom. He went on to sell 12 million cans, making $7 million (over £5 million). People were gasping for cleaner air and willing to pay for it.

The Airpocalypse wasn't, in reality, a single air pollution event. It was simply the worst of many that happened like clockwork every winter. In the 30 years to 2015, lung cancer rates in China increased by 465 per cent, during a period when tobacco smoking rates had actually gone down. In 2013, an 8-year-old girl in Jiangsu became China's youngest lung cancer patient; the doctor treating her believed it was caused by air pollution exposure. During the 2015 Beijing Marathon, six people suffered heart attacks as a result of PM2.5 levels that day. A journal paper that year found that on Beijing's worst days, an ordinary citizen would breathe air pollution equivalent to smoking 25 cigarettes. The worst day recorded in 2015 in the Chinese city of Shenyang saw PM pollution reaching 1,400µg/m^3 (almost three 'crazy bad' days rolled into one) – equivalent to smoking 64 cigarettes. Even the most hardened chain-smoker would struggle to smoke that many.[*] The difference of course was that in

---

[*] Given an average of 10 minutes per cigarette, it would take 10.5 hours to smoke 64, one immediately after the other.

Shenyang, the city's infants and infirm were effectively chain-smoking too.

China's smog had managed to encapsulate both London's past and present predicaments: domestic coal and solid fuel smoke, industrial pollution and modern transportation fumes. China only had 18 million vehicles on its roads as recently as 2001; by 2015, vehicle ownership had reached 279 million.

Delhi, 2017

On 6 November 2017, Belgium's King Philippe and Queen Mathilde attended a ceremony in Delhi. Commemorating Indian soldiers who fought in Flanders during the First World War, a guard of honour was shrouded in a haze reminiscent of the smoke from artillery fire and mustard gas exactly a century before. However, this wasn't planned as a re-enactment. At 11 a.m. that day, PM2.5 levels at the US embassy in Delhi reached an asphyxiating $986\mu g/m^3$. When Prime Minister Modi posed with the Belgian royal couple for a group photograph, they could no longer see the guard of honour. The next day, PM levels peaked at $1,486\mu g/m^3$ – amongst the highest ever recorded. The level wouldn't dip below triple figures until 17 November (and even then, only briefly).

According to the Delhi-based non-profit Centre for Science and Environment, 'the key contributors to this smog in Delhi and its vicinity were vehicles; unchecked construction and road dust; garbage burning; burning of paddy residues by farmers in Punjab, Haryana ... near-still weather conditions without wind; the onset of winter; and of course, the Diwali firecrackers'. In response, Arvind Kejriwal, the city's chief minister, ordered schools to close for three days, halted construction and demolition work for five days and shut down the city's central Badarpur coal plant for 10 days. The international press ran headlines such as '"I feel helpless": Delhi residents on the smog crisis' (*Guardian*, 8 November

2017) and 'Air quality in New Delhi "worse than smoking 50 cigarettes a day"' (*Sky News*, 11 November 2017).

On 19 November, the Indian TV channel NDTV opened its debate programme with the following impassioned intro: 'This week, if you live in North India you probably wish you didn't as a blanket of smog-poisoned toxic air descended over city after city. Yet life seems to continue as usual. When air quality levels go from severe to very poor we actually cheer … there's silence from the very top with the environment ministers saying "at least it's not as bad as the Popol gas tragedy", even as study after study revealed how millions of Indians are dying of this pollution. How can we as citizens, and government health care professionals, come together to save ourselves from this health emergency?'[4]

On 22 November, I arrive in Delhi. As my flight begins its descent, a blanket of grey cloud awaits below for the plane's wings to slice through. As we get lower, however, the grey looks too flat, too uniform to be cloud. And it is translucent – the buildings can be seen through it, like pebbles in the bottom of a muddy puddle. It dawns on me that it is a sunny, cloudless day: this is the smog.

At the airport I seek out a local SIM card for my phone and order a taxi – an ingrained guilt nags at me for not using public transport, but the car is integral to the story of modern Delhi. I need to see the roads for myself. The traffic moves in a stop-start stutter, rarely less than four or five rows deep, irrespective of how many lanes there actually are, each driver trying to fill the smallest gaps that appear, seeking an imaginary advantage. On the back seat of the cab, my Laser Egg shows PM2.5 above $300\mu g/m^3$: levels I have never personally experienced before. I make a mental note to do everything as slowly as possible to keep my breathing and heart-rate down. In the coming days, I become familiar with a constant sound on the streets of Delhi: a hacking clearing of catarrh. Drivers and pedestrians alike are forced to clear their mucus and spit it out onto the ground. It would seem rude if it wasn't so necessary. At points during the trip I find myself doing so, too.

At my B&B, my host Vandana welcomes me with toast and hot tea (she clearly knows the British well). 'What a great day to have arrived,' she exclaims. 'The smog has gone!' 'Gone?' I reply, taken aback, given the milky gloom in the sky and my Egg readings. 'Oh yes. Last week it was around 1,000,' she says, referring to the AQI.* 'It's now only 200 or so. It lifted a day or so ago.' Still she advises I buy a facemask, if only for when I walk beside the roads. 'Your throat will probably feel sore at the end of the day,' she says, 'but if it does – just warm some water in the kettle and gargle it. It's like taking a bath. It usually does the trick just fine.' When she leaves me to unpack, I reach for the indoor air purifier in the room and turn it up a notch. But as the curtains flutter around loose-fitting window frames, I know it won't do much good. When I go to bed, my Egg never dips below $70\mu g/m^3$. The next morning, a handful of labourers are demolishing the block of flats opposite my apartment, by hand, brick by brick. They light a fire to keep warm and cook on. I can't see what they are burning but it smells sickly sweet. My Egg reading rises above $200\mu g/m^3$.

Later, at the Indian Institute of Technology [IIT], large expanses of lawn and private roads loop around 1970s concrete department blocks, few more than three stories, spanning the immediate horizon. On the second floor of the Civil Engineering block I look for the brown door with the nameplate 'Professor Mukesh Khare'.

'I joined IIT Delhi as a professor in 1991, and since then I have been working in the area of air quality,' Prof Khare tells me. 'Delhi used to have a problem with CO [carbon monoxide] in the 1990s. But … the problem now is NO_2 and CO, because of the high temperature fuel, and also PM2.5.

---

* The AQI – Air Quality Index – is used by some government agencies and air quality apps to bundle all pollutants together and communicate an overall health warning. But as a rough rule of thumb, when an AQI gets above 200, the PM2.5$\mu g/m^3$ will be a similar number.

Diesel is another contributor.' Delhi was recently ranked the most polluted major city in the world for ambient air pollution. The official annual mean PM2.5 level in Delhi in 2014 was 153μg/m³, 15 times over the WHO guideline, with daily levels often exceeding 550μg/m³ (above Beijing's 'crazy bad' level of 500).

When I visit the Central Road Research Institute (CRRI), Dr Niraj Sharma, senior principal environmental scientist, who has studied road emissions for the past 25 years, reaches into his desk drawer. 'I am very fond of collecting the newspaper cuttings,' he explains. 'You take a glance at these.' He pulls out a thick pile of newspaper articles that he has carefully cut out. I read some of them aloud: 'Air quality improves slightly in Delhi, falls to very poor again', 'Less pollution in Delhi this Diwali, but Air Quality still bad, way beyond safe levels', 'Government says air pollution spiked after fireworks started', 'Low wind speed keeps Delhi air very poor', 'Delhi, you're killing me', 'Smog chokes city, doctors declare health emergency'. Would you agree with that, I ask – is it a health emergency? 'Yes. This time I will say. For the first time in 25 years, around 10 days ago, I felt sort of suffocation outdoors … I've felt discomfort before many times, but now I felt choking.'

The population of Delhi, the second-largest city in the world at the time of writing, is forecast by the UN to grow from 24.9 million in 2014 to 36 million in 2030. As the city expands, so does the number of cars and roads. Dr Sharma tells me that in 2010 the number of vehicles in Delhi was approximately 6 million; 'Now it is 10 million.' Vehicle pollution has been found to account for around 72 per cent of total air pollution in Delhi, compared to just 23 per cent in 1970–71. Other Indian cities didn't choose the same route to development. Greater Mumbai only saw an increase from 1 million to 1.7 million vehicles over the same period, despite having a similarly large population of 20.7 million.

Shubhani, a Delhi businesswoman, tells me, 'Every year you see that kids are going on to nebulisers, they are coughing, going off sick – especially the young ones. One of my friends has her kid perpetually on some kind of medication. And there is really nothing else as a cause, other than the air that we breathe.' Keen to show me some evidence, she searches for an image she saved on her phone. It is a newspaper front page from a couple of months previously, showing the leading causes of death in India: in 1990, the top five were diarrhoea, lower respiratory complications, pre-term birth complications, TB and measles. By 2016, heart disease was suddenly top of the list, making up 8.7 per cent of the total disease burden, compared to only 3.7 per cent in 1990.[5] Second came chronic obstruction of lung, then again diarrhoea, then lower respiratory complications, then stroke. Every new entrant in the top five had a strong air pollution link.* The head of the air laboratory in the Central Pollution Control Board told the *Hindustan Times* in July 2017 that Delhi hadn't experienced a single day in the previous 535 days in which air quality could be rated 'good'.

Jyoti Pande Lavakare, an economist, financial journalist and self-described 'Delhi high-school Mom', is co-founder of Care For Air, a clean air campaign group. At her house in a leafy, middle-class neighbourhood, public parks appear between every other block, watered excessively. Guards sit outside each entrance gate. Such neighbourhoods are not supposed to be infected by something as unpleasant as pollution, yet my Egg tells me it is $280\mu g/m^3$. Shards of sunlight illuminate the outdoor air like dust in an attic. Jyoti tells me that some foreign embassies now categorise Delhi as a 'hardship posting' – they don't send diplomats with families to Delhi any more. United Airlines recently cancelled flights into Delhi during the worst of the November smog. 'It has been building among the

---

* As we'll see in Chapter 6, stroke and heart disease combined cause more air-pollution-related deaths than lung conditions.

international community as an open secret,' she says. 'It became something everybody knew but didn't want to discuss.' Her phone rings loudly to the ringtone of Coldplay's 'Yellow'. 'I will take it if it is the doctors,' she says apologetically. 'It's OK, it's not … That is the funny thing, Tim. I am now fighting this fight at a more personal level. My mum has just been diagnosed with lung cancer. She is a long-time Delhi citizen. We have no history of cancer in our family … she is not a smoker. She just breathes the Delhi air.'

Just days after I leave Delhi, an international cricket match held in the capital – a grudge match between India and Sri Lanka – saw smog levels such that umpires halted play for 20 minutes to consult with team doctors. It was the first recorded instance of an international cricket match being halted due to smog. The match resumed but two bowlers left the field complaining of breathing difficulties. 'We had players coming off the field and vomiting,' the Sri Lanka coach complained to reporters. 'There were oxygen cylinders in the changing room.'[6]

LA's 'Photochemical Smog'

The first smog to descend on Los Angeles arrived in 1943. Many feared it was a Japanese chemical attack, causing panic on the streets. But in fact, the menace was home-grown, and its arrival was predictable. Los Angeles, like Mexico City and Beijing, is a natural pollution trap. It sits in a geological bowl (of around 1,630 square miles) surrounded by a high rim of mountains which traps in the air below. Early in the twentieth century, fumes from steel mills, chemical plants and waste incinerators slowly began to fill up the bowl. After the 1943 smog, government agencies turned to Dr Arie Haagen-Smit, a bio-chemist at the California Institute of Technology, to investigate what was happening. His subsequent research discovered the process that causes ozone pollution and coined the phrase 'photochemical smog'.

Haagen-Smit's paper 'The control of air pollution in Los Angeles', published in the December 1954 edition of

Engineering and Science, concluded that: 'The chemical analysis of smog air shows the presence of a great number of materials, among which are sulfur dioxide and dusts, well known as trouble-makers in other industrial areas ... our discovery that photochemical oxidation of organic material is accompanied by ozone formation gives further scientific evidence for the necessity of controlling hydrocarbon emissions.' Haagen-Smit's obituary in the *New York Times*, 1977, remembers: 'he engaged the oil and automotive industries in an almost single-handed battle against pollution from burned fuels ... induced industry to filter smokestack fumes, urged automotive plants to develop hardware that would reduce exhaust vapors and reiterated the need for planned control of industrial expansion.' One thing he couldn't stop, however, was the city's love affair with the car. Dr Devra Davis estimates that by 1955, around half the city's five million residents had a car, burning a total of 58,000 tonnes of fuel a year. Los Angeles was previously known for its electric streetcars, covering 1,500 miles of track, but – as in London – they were ripped up to make way for the automobile. Davis writes that, in 1954, 'huge bonfires were lit as kerosene-soaked streetcars and electric trains that formerly served Hollywood were burned.'[7] If ever an image summed up the mid-twentieth-century mistake to favour fossil fuel over electrification, it is surely this.*

From 1950 to 1970, the population of southern California doubled, while the number of vehicles tripled.

---

* In 1900, there were 600 electric taxis driving around New York, accounting for around a third of all vehicles on the road. Even Porsche produced the all-electric model 'P1' in 1898. So, what happened? In 1908 the first affordable mass-produced car, the Ford Model T, chose the rival gasoline-powered engine. Then crude oil was discovered in Texas, suddenly flooding the market. Pretty soon, world power rested on the control and distribution of oil, and greater consumption was encouraged and subsidised. Electric transport faded as a minor domestic concern by comparison.

Governor Ronald Reagan urged residents in 1974 to limit 'all but absolutely necessary auto travel' and to drive slower to reduce emissions, but his words had little effect. Studies in California in the 1980s began examining the health effects of traffic pollution and found an unexpectedly high rate of premature deaths from cardiovascular disease.

Mary Nichols, chair of the California Air Resources Board from 1979 to 1983, recalls, 'just reducing the amount of [volatile organic compounds, or VOCs] in the air ... was very controversial, and it was litigated. It meant hours in hearings with companies like Southern California Edison which fought a vigorous battle with all kinds of science experts trying to convince us that we shouldn't be trying to curb emissions.' At its worst, the LA smog 'burned your lungs when you breathed it, and it smells bad – it literally has a smell and a taste to it – which is not what air is supposed to be like,' says Nichols. 'It has an industrial, chemical flavour. And it's also ugly – you can see it, it casts a pall over everything you are looking at. And you can see the effects of it, on buildings, on monuments, where it eats away at marble and stone. And as far as humans are concerned and animals – because pets experience it too – it hurts to breathe and makes people less capable of being outdoors. People would be told not to go outside ... they were stuck indoors.'

Sam Atwood was a daily newspaper reporter in the 1980s at the San Bernardino *Sun* and the Santa Fe *New Mexican*. In 1990 he won a national journalism award for an eight-part series on the health effects of smog. 'Around 1987 the opportunity came up to work for a newspaper in Southern California,' he remembers, 'so I flew out for an interview – and you can still have this experience today ... you know, you are literally flying into a blanket of smog ... The San Gabriel mountains are beautiful, they're less than 10 miles from the airport but you couldn't see them at all ... I almost started having a little bit of a panic attack thinking "What am I doing here ... I'm going to be breathing this stuff?!"' Sam's plane landed during one of the many frequent Stage 1 smog alerts, signalling a high level of ozone: 'at that level anyone would feel the effects, it's difficult to take a deep breath, you feel like there's a heavy weight on your chest and there's no denying the fact that air quality is affecting each and every person who is breathing it.' Los Angeles, he says, 'has always had a horrific ozone problem and it continues to be our greatest challenge'.

Between 2012 and 2014, LA had 81 'red alert' high ozone days; by contrast, the entire state of Florida had just one. Suzanne Paulson, UCLA (University of California, Los Angeles) professor and director of the Center for Clean Air, explains that the problem today is 'still internal combustion cars plus other sources like off-road vehicles, construction vehicles, planes, trains, and ships: we have an enormous amount of shipping coming into LA, about half of all the stuff that comes from Asia to the United States comes in through the ports of LA and Long Beach.' In summary, says Paulson, 'if we suddenly got rid of all the combustion then we would pretty much deal with all of the air pollution.'

The 2016 State of the Air report by the American Lung Association found that 12 cities in California reported a year-on-year increase in ozone pollution, LA having topped the 'most polluted' list for 15 out of the last 16 reports. During 2016, five cities in the Western states had their worst short-term daily pollution episodes since the report began, largely due to an increase in summer droughts and wildfires, 'although they are not the only sources. High particle days frequently result from use of wood-burning stoves for heat, dust storms, wildfires and weather patterns that trap in emissions from power plants, trucks, buses, trains, ships and industrial sources.' Across the nation, more than 4 in 10 Americans (44 per cent) live in counties that have unhealthy levels of ozone or particle pollution, a higher percentage than in 2009–11. PM2.5 alone causes twice as many deaths per year than traffic accidents in the US. The American Lung Association also warned of powerful forces seeking 'to weaken the [US] Clean Air Act* ... and to undermine the ability of the nation to fight for healthy air'. With the incoming Trump administration, they would very soon be proved right.

Paris: NOx-ious gases

When I arrive at the Airparif offices on the south bank of the Seine, Amélie Fritz apologies for looking tired. It is the day after the city's '*Journée sans Voiture*', a now annual day when Paris tries to persuade its residents to travel without a car for a day. Fritz, an environmental biologist at Airparif, the city's official air quality monitoring organisation, was talking to journalists until late into the night. She did the last TV interview at her home – 'I was too tired to stay at work, so I just said "yeah OK sure, but you have to come to my house now".'

She shows me upstairs to her office, where a large desk spills over with various reports and journal papers. 'Paris has quite good luck in terms of geographical conditions,' she begins. 'The land is very flat, we are not surrounded by mountains, there is no massive industry, it's not an industrial site, there's lots of wind and rain – this is quite good for air quality ... But if it is low sky, no wind – it's like running your car inside your garage, basically.'

From 30 November to 17 December 2016, the Greater Paris area experienced one of its longest and most intense pollution episodes in a decade. A high PM10 concentration due to local emissions from traffic and domestic fires, plus no wind, saw the whole of Paris trapped in its garage with the car

---

* The Clean Air Act of 1970, amended in 1990, gives the US Environmental Protection Agency (EPA) the mandate to regulate air pollution emissions and set National Ambient Air Quality Standards (NAAQS) – or limits – for six major 'criteria air pollutants' considered harmful to public health: nitrogen dioxide, particulate matter, sulphur dioxide, carbon monoxide, lead and ozone – with states required to produce State Implementation Plans showing how they intend to meet the targets.

running. On Wednesday 30 November the hourly PM2.5 reading in Saint-Denis, central Paris, peaked at $195\mu g/m^3$, while NO_2 maxed out at an asphyxiating $283\mu g/m^3$ at Place de l'Opera the following day. The situation improved during the weekend, but then went back up again the following week. All public transport plus the city's public bicycle-hire and electric car-hire schemes were offered free of charge to try to stop residents from firing up their car engines. Over 2,000 asthmatic children were treated in Parisian hospital emergency rooms. By 8 December, the front page of the daily newspaper *20 Minutes* exclaimed '*La fumée tue*' (Smoke kills). TV news channel *France 24* commented that 'the local authorities and the government are at loggerheads with each other, blaming each other for these current pollution levels', while the campaign website 'Stoppollution' claimed, 'Living in Paris during this peak of pollution is equivalent to breathing the smoke of eight cigarettes a day in a room of 20 square metres.'

According to the national public health agency Santé publique France, air pollution kills 48,000 people a year in France, 34,000 of which are classed as 'avoidable' deaths: 'In urban areas with more than 100,000 inhabitants the results show, on average, a loss of 15 months of life expectancy at 30 years due to PM2.5; in areas between 2,000 and 100,000, the loss of life expectancy is 10 months on average; in rural areas, on average, nine months of life expectancy are estimated to be lost.' But crucially these findings aren't related simply to peak episodes such as December 2016. A Santé publique France study of 17 cities in France from 2007 to 2010 confirmed, 'it is the daily and long-term exposure to pollution that has the greatest impact on health, with pollution peaks having a marginal effect.'

Many of the problems began in the 1970s. Les Halles market was demolished to make way for wider roads along the banks of the Seine and an underground road tunnel was built to make central Paris more accessible to cars. The city was then encircled by a ring road, the Boulevard Périphérique, completed in 1973. Rather than ease congestion, this simply attracted more traffic. In 2010, around 3.6 million Île-de-France residents (people living in the region around Paris) were potentially exposed to traffic-derived NO_2 levels exceeding the annual limit value, while roadside levels of NO_2 had increased every year since 1997. While the PM2.5 problem in Paris may seem slight compared to Delhi or Beijing, its NOx problem (NO_2 + NO) is worse than both cities. The 2010 Airparif annual report explains, 'the evolution of nitrogen dioxide levels, both at background and at roadside sites, probably relates to primary NO_2 emissions from diesel-powered vehicles. Although the filters that now equip most new diesel vehicles contribute to reducing particulate emissions, they also give rise to a significant increase in NO_2 emissions. It is now confirmed that the proportion of NO_2 in NOx emissions is increasing steadily.' The warning procedure alerting Parisians of high pollution levels was triggered on 44 separate days in 2012. By 2013, there were 14.6 million trips taken by car in the Île-de-France region every single day, an estimated 65 per cent of which took place in Paris.[8] The 2017 Airparif report finds that NO_2 levels along major roads are twice that of those away from the road, and often two times higher than the annual EU limit.

Amélie describes the air pollution in Paris now as being a combination of 'NO_2 plus PM10 and PM2.5. There are still some remaining issues with benzene. And ozone outside of the city, on the regional level … There is also quite a lot of agriculture and wood smoke as well.' She tells me that air pollution is now 'one of the major public preoccupations – it's actually coming second only to work and jobs. So, it's a massive concern basically, people are very worried for their health.' Polling conducted by Airparif confirms that half of all Paris residents are concerned by the NO_2 levels exceeding health-based thresholds.

In 2017, Parisian Clotilde Nonnez, a 56-year-old yoga teacher, took the unprecedented step of taking the French state to court for failing to protect her from the effects of air pollution. Having lived in Paris for 30 years, she had seen her health deteriorate despite a very healthy diet and exercise. When it became worse than ever during the December 2016 smog, she suspected a link. Her doctors confirmed it. 'The doctor treating me says Paris air is so polluted that we're breathing rotten air,' she told *France Info*. 'She has other patients like me, including children and babies too. My cardiologist says the same.' Nonnez's lawyer François Lafforgue told *Le Monde* newspaper that something needed to be done to stop 48,000 French deaths per year: 'We are taking the state to task because we think the medical problems that pollution victims suffer are as a result of the authorities' lack of action in tackling air pollution.'

On 15 February 2017, the European Commission sent a final warning to France for failing to address persistent breaches of NO_2 limit values in 19 air quality zones, including Paris. It stated, 'EU legislation on ambient air quality and cleaner air for Europe (Directive 2008/50/EC) sets air quality limits that cannot be exceeded anywhere in the EU, and obliges Member States to limit the exposure of citizens to harmful air pollutants. Despite this obligation, air quality has remained a problem in Paris … Of the total emitted NOx from traffic, around 80 per cent comes from diesel powered vehicles.'

The Global Air Con

The five cities mentioned above offer just a snapshot of what, by the 2010s, had become a global problem. Air pollution has overtaken poor sanitation and dirty water to become the number one environmental cause of premature death in the world. The latest estimate from the WHO is that approximately 4.2 million people die from outdoor air pollution annually, far greater than the number from HIV/AIDS, tuberculosis and car crashes combined. According to WHO figures in 2018, nine out of ten people around the globe now breathe air containing high levels of pollutants. Unicef believes that two billion children live in areas where pollution levels exceed the WHO air quality standards, while nearly 600,000 children under the age of five die annually from diseases caused or exacerbated by air pollution.

This clearly isn't a problem that is limited to Europe, India, China or the US. According to the WHO ambient air pollution database, the most polluted city in the world in 2016 was Zabol in Iran. Representing Africa in the 'top 50 most polluted' are Bamenda in Cameroon, Kampala in Uganda and Kaduna in Nigeria (although the WHO is keen to point out that 'Africa and some of the Western Pacific have a serious lack of air pollution data'). The high altitude of South American cities such as Bogotá, Colombia, means they are choked with diesel pollution trapped within mountain basins. In fact, almost all cities – 97 per cent – in low and middle-income countries do not meet WHO air quality guidelines.

The whole world has a smoke problem. If it's combustible – and especially if it's a fossil fuel – we'll happily burn it, with scant regard for what's in that smoke or where it ends up. So, what *is* in that smoke? And where *does* it end up?

CHAPTER TWO

Life's a Gas

It quickly became obvious that I needed a crash course in chemistry – specifically atmospheric chemistry. Some pollutants are worse than others, some sources more significant than others, and I needed help distinguishing between them. My physical journey towards scientific enlightenment began much like my mental one: I was lost, wandering aimlessly around a nondescript grey roundabout on a nondescript grey day. This particular roundabout (the actual one, not the mental one) happened to be in the outskirts of York, England. Fortunately, Professor Ally Lewis bounded out to find me. The university summer term was over, and the normally busy, buzzing campus has the feel of an out-of-town retail park after the shops have closed. A few international students stroll lazily around and post-docs boost the average age, while academics like Ally Lewis make the most of the peace and quiet to work on journal papers and funding applications. Despite being a professor of atmospheric chemistry with decades of experience in the field, Ally looks as un-professor-like as possible, dressed in jeans and a faded Patagonia T-shirt.

Fittingly for a chemistry department, as we approach the entrance a small truck is refilling canisters of liquid nitrogen directly outside the doors. Unnerved, I swerve quickly past.* 'We used to have local school teachers come by and borrow a bit for their lessons,' Ally tells me. 'They'd bring a thermos flask and we'd let them fill one up for free. You can't do that now, of course,' he adds, though he's not sure exactly why not.

---

* I've clearly watched too many sci-fi films where a body gets frozen in liquid nitrogen, before inexplicably falling to the floor and shattering into thousands of pieces.

He shows me around the labs at Wolfson Atmospheric Chemistry Laboratories, a new building (built in 2012) that wouldn't be here if it weren't for the renewed interest in air pollution. It was never a sexy enough subject to attract benevolent funding before. In the early 2000s, atmospheric chemists such as Ally were barely clinging on. His department used to rough it with several others in an old 1960s building, with labs that would echo with noisy analysis machines. In the labs he shows me now, thanks to soundproofing panels that hang from the ceilings and clad the walls, the same machines make little more than a background hum. A mass spectrometer runs analysis of aerosols from Malaysia. A gas-measurement machine the size of a car engine sits gleaming on the floor, fresh from a flight measuring high-altitude pollution. In a very British way, alongside this expensive cutting-edge equipment sit more amateur-looking set-ups – tinfoil wraps and bottles bubbling with looping copper wires.

Ally stops by a huge picture of a monitoring station in Cape Verde, where York University has studied transatlantic pollution for decades. Last year, he says, it recorded the first increase in ethane levels for 40 years, probably due to the US fracking industry. But you can't understand one gas, or one pollutant, in isolation, he says. There is complex chemistry going on. 'We have an atmosphere that is essentially low-temperature combustion – it is a bit like a fire and it's continually burning up most of the crap that we put into it. If we didn't have a low-temperature combustion atmosphere, the concentrations of pollutants would just build up and build up. So, we rely on the fact that virtually everything we emit in the end is reacted away to become CO_2 and water.' Virtually everything? 'Yes, not everything.'

When it comes to gases, there are some lead actors in air pollution, and some notable others with walk-on parts (albeit ones that sometimes steal the show). Clean air is made up of 78.09 per cent molecular nitrogen, or N_2. The rest is 20.95 per cent oxygen, plus trace gases such as 0.93 per cent argon and 0.04 per cent carbon dioxide. Life on Earth has grown accustomed to (or, to put it another way, reliant upon) that delicate balance. In polluted air, by definition, we are breathing in something that we shouldn't. Something has muscled its way alongside the trace gases that our bodies are not designed to cope with. So, who are these interlopers in our air, how did they get there, and which should we be most concerned about?

The leading actors

Nitrogen dioxide

The nitrogen that we breathe (called 'N_2' because it is two atoms of N – nitrogen – bonded together) is inert, meaning that it doesn't react chemically with anything else. Think of it as the cocktail mixer that carries the good stuff, in this case oxygen, to our bloodstream.[*] But so-called 'reactive nitrogen' is chemically different because something has split the molecules of nitrogen apart and bonded them to something else. Sometimes this happens naturally, such as bolts of lightning, which have enough heat to bind nitrogen and oxygen together to form nitrogen oxide gas. But in our towns and cities, the biggest single contributor of the nitrogen oxides called nitrogen monoxide (NO) and nitrogen dioxide (NO_2) – collectively referred to as 'NOx' – is transport fumes. Every car engine is, in effect, a mini lightning bolt, and our roads are filled with a never-ending storm. Nitrogen dioxide can cause significant health problems, and is therefore of the greatest concern, while nitrogen monoxide is less harmful but tends to react quickly in the air to form more NO_2.

In the UK, a Department for Environment, Food and Rural Affairs (Defra) study in 2015 showed that around 80 per cent of urban NOx emissions come from transport, and one-third of those are just from diesel cars. Gas combustion from boilers makes up much of the remaining 20 per cent. Aviation also scatters some NOx down on us from above, accounting for 14 per cent of all transport NOx emissions in the EU. NOx

---

[*] Nitrogen is also a relatively heavy gas, meaning that over millennia the Earth's gravity has held on to it, giving us the atmosphere we have now. When you get to the far reaches of our atmosphere – the thermosphere – the lighter gases, helium and hydrogen, are more abundant.

emissions from air travel in Europe have doubled between 1990 and 2014 and are forecast to grow by a further 43 per cent between 2014 and 2035. In 2010, a computer model of flight-path records uncovered a movement of pollution that spread not just across national borders but from one continent to another. Planes flying at cruising altitude were found to distribute their fumes, carried by wind currents, as far away as 10,000km (6,200 miles), typically to the east of the plane's route. Flights over Europe and North America at high altitude therefore tend to blow across Asia. Meanwhile, industry and energy production typically accounts for around a quarter of total NOx pollution. As for where it ends up, it inflames our lungs and acidifies soils, forests and water. Algal blooms caused by reactive nitrogen in the Gulf of Mexico were first discovered in the 1950s, causing oxygen-free dead zones as the build up of dead algae – which fed on the nitrogen pollution – became too intense. In 2017, the dead zone was found to have grown to $22,720km^2$ (by way of contrast, the famous 'Dead Sea', which borders Israel and Jordan, is just $605km^2$).

Ammonia

Another reactive nitrogen-based gas in the air is ammonia (NH_3), caused by a single atom of nitrogen bonding with three atoms of hydrogen. In the atmosphere, ammonia can be breathed in and irritate the lungs and eyes, and does an even better job than NOx at degrading ecosystems. According to the UN Food and Agriculture Organization (FAO), agriculture is the dominant anthropogenic (human-produced) source of ammonia. Cattle and poultry, to name just two, are fed nitrogen-rich protein feed, much of which is not digested but comes out in the urine and manure, releasing ammonia into the air. 'One of the chemical peculiarities of air pollution chemistry,' says Ally, is that 'ammonia and NOx are individually both gases yet when they react together in air they form tiny liquid droplets that can be inhaled. If those tiny droplets grow in just the right environment they can even go on and seed clouds, and change how much rain falls. We are only now appreciating just how much pollutants can seep into other areas of the environment, and even affect the weather.' Global ammonia emissions in pre-industrial times are thought to be less than 30 per cent of present levels, due to the world's ever-increasing population and appetite for meat. In the EU, 94 per cent of ammonia emissions came from agriculture in 2015, half of which were from cattle, and a quarter from fertilisers.

Ozone

But while NOx is the current star of air pollution and attracts most of the press attention, ozone has been making headlines for even longer. In LA in the 1950s, Dr Haagen-Smit of the California Institute of Technology discovered that ozone is 'produced in the air by the action of sunlight on organic material in the presence of nitrogen dioxide'. In other words, when nitrogen dioxide hangs in the air alongside particles of organic matter (VOCs – which we'll meet on the next page) then, if it's a sunny day, you get ozone. This can happen in a matter of hours, especially on hot cloudless days, with outbreaks leading to peaks in hospital admissions. Ozone exposure inflames lung tissues and triggers asthma attacks, with children and the elderly being most at risk. According to the European Environment Agency, ozone exposure today causes 14,400 premature deaths in Europe annually. But this isn't just a problem for us. High levels of ozone are toxic to most plants and animals too. Major crops such as wheat, maize, rice and soybean are sensitive to surface ozone pollution, leading to concerns for global food security.

The current health-based standard in the US for ozone levels is 75 parts per billion (ppb). Because milligrams and kilograms are units of mass they don't work as well for gases, which tend to be referred to in terms of concentration – how many parts you'd find within a million or a billion molecules of air. A useful little factsheet from the US National Environmental Services Center (NESC) suggests that one part per billion is like adding a pinch of salt to a 10-ton bag of potato crisps (OK, they said 'chips', but come on – I'm British). If the gas pollutant is found in abundance – meaning you start adding the salt by the bucket load – then it's easier to zoom in and measure in parts per million or billion instead. The current health-based standard in the US for ozone levels is 75 parts per billion (ppb). The WHO reduced its recommended limit for ozone in 2016, from the previous level of 60ppb to a 50ppb eight-hour mean, based on what it called 'recent conclusive associations between daily mortality and lower ozone concentrations'. At the time Haagen-Smit was making his discoveries, LA's ozone exceeded 600ppb.

Whereas most other air pollutants are worse the closer to the source you get, outbreaks of ozone typically occur many miles away from the source. 'People tend to think of regional ozone forming over two to 24 hours,' says Ally. 'If you think an average wind speed is 10 metres per second, you're making it over the next 200 to 400 kilometres of the passage of that air. So, it's generally not an issue right in a city centre, but the ozone slowly increases as you move away.' That's also why cities like LA, Beijing and Bogota are in a double-bind: in a basin surrounded by mountains, the ozone is trapped and stews like a forgotten tea bag.

Ozone's reaction with the sun's ultraviolet rays (think again of the atmosphere as a flame, constantly burning) is also the main source of hydroxyl (OH), a single oxygen atom bonded with a single hydrogen atom. This creates a 'free radical', which is highly reactive, very short-lived, but still able to do a lot of damage in that short space of time (the atmospheric lifetime of hydroxyl ranges between 0.01 and 1 seconds). Suzanne Paulson, professor of atmospheric chemistry at UCLA, describes OH as 'a gnarly little molecule that really just wants to pull a hydrogen off of something and become water ... and it pretty much always wins the battle.' As a result, it oxidises (corrodes) almost everything in its path.* But hydroxyl isn't entirely a baddie. It does a vital clean-up job in the atmosphere. Without hydroxyl, says Paulson, the greenhouse gases in the atmosphere would simply build up and up in our atmosphere. Think of hydroxyl, then, as an ultra-aggressive guard dog; it's effective, but you don't want to let it off the leash.

---

* I look at the health effects caused by 'oxidative stress' and free radicals in chapter 6.

Ozone is of course famous for another reason: the 'ozone layer', which occurs naturally in the stratosphere, around 9 to 22 miles (15 to 35 kilometres) above the Earth's surface, and forms a protective layer that shields the Earth from the sun's UVB rays. Over 91 per cent of the ozone in Earth's atmosphere is in the ozone layer, essentially allowing life on Earth to exist – it is the planetary equivalent to the sun cream we slather on our skin. So, ozone up there is good. We just don't want it at ground level. And that remaining 9 per cent of ozone is produced at ground level almost entirely due to our own emissions.

Volatile organic compounds

The third essential precursor of ozone, alongside sunlight and NO_2, are VOCs – volatile organic compounds. They are 'volatile' because they have low boiling points, meaning they evaporate easily into the air. While 'organic' doesn't mean you can buy them at the farmers' market; the chemistry definition of 'organic' means it has both carbon and hydrogen atoms. And 'compounds' because there is a myriad of different forms, depending on the number of rings of carbon and hydrogen atoms they have. When I speak to Ally in his small office, he points out, 'There's probably 3,000 different VOCs in this room now.' Gloss paint, for example, gives off xylenes. The smell at the petrol station when you fill up your car is a mixture of benzene (C_6H_6), toluene (C_7H_8) and xylene (C_8H_{10}). Globally, the majority of VOCs come from trees, largely in the form of isoprene (C_5H_8). The release of VOCs gives the blue-ish haze above forests, such as the Smoky Mountains in the US, or the Blue Mountains in Australia. This led president Ronald Reagan to say in 1981 that 'Trees cause more pollution than automobiles do,'* which in one very basic sense was right: 'Ninety per cent of VOCs come

---

* Reagan is also reported to have told an audience of enthusiastic lumberjacks at the Western Wood Products Association, in 1965: 'A tree's a tree. How many more do you need to look at?'

from trees, that is actually true,' says Ally, 'but of course the vast majority of them are being emitted in the Amazon where it's completely clean, there's only the VOC, there isn't any NOx, so they go up into the air, they get oxidised and it just disappears, it's a natural cycle.' Whereas in a city, he says, 'If you had 20ppb of NOX and 1,000ppb of VOC, and the sun was shining, you would be making ozone at a rate of about 20ppb per hour, so a really high rate.'

VOCs are mostly of interest, then, because they help to form ozone. But some VOCs are directly harmful. Benzene is a known carcinogen while toluene is toxic for the central nervous system. A study in the Nigerian megacity Lagos in 1995 found outdoor benzene values of $250-500\mu g/m^3$ – the legal limit in the EU is just $5\mu g/m^3$. Some sources of benzene and toluene are relatively obscure, such as inks, cleaning agents and nail polish. But the main one comes as little surprise: vehicle fuel. The 1995 Lagos study pinpointed the pollution as 'caused by the combination of many strongly emitting vehicles and frequent traffic jams' and 'diesel vehicles emitting plumes of black smoke'.[1]

Some of the more complex, and the most dangerous, organic compounds are the polycyclic aromatic hydrocarbons (PAHs), such as benzo[c]phenanthrene $(C_{18}H_{12})$, benzo[a]-pyrene $(C_{20}H_{12})$ and benzo[e]pyrene $(C_{20}H_{12})$. These are formed by burning fossil fuels and biomass (organic fuel that hasn't yet been fossilised, such as agricultural residue). PAH exposure has been linked to respiratory problems and cancer, causing genetic mutations. Naphthalene $(C_{10}H_8)$, which due to its pungent aroma is the primary ingredient in some mothballs, can cause the breakdown of red blood cells if inhaled or ingested in large amounts. At the time of writing, I could – but didn't – buy a 100g bag of 'naphthalene balls' boasting '99 per cent purity' on eBay for £3.32 (US $4.65).

One of the biggest sources of the PAH benzo[a]pyrene is agricultural stubble burning – the practice of burning crop stubble in the fields after harvest. In Rwanda, Africa's second most densely populated country, most of the population works in subsistence agriculture. The haze from agricultural burning in Rwanda can go on for half the year. Dr

Langley DeWitt, Station Chief Scientist for the Rwanda Climate Observatory, a joint MIT-Rwanda project, explains that 'the observatory is on a rural mountaintop and we measure air masses from central, eastern, and southern Africa … What we are finding is high influence of biomass burning seasons in Rwanda … Both burn seasons take place during Rwanda's two dry seasons in December–February … and June–August.'

Agricultural fires are also lit to clear new land, including the burning of pristine forest. Between 1976 and 2010, more than 750,000 km^2 of the Brazilian Amazon was deforested, equivalent to 15 per cent of the original rainforest, causing over half of Brazil's PM2.5. In equatorial Asia, fires are commonly used to clear scrub vegetation and peatland areas. In September and October 2015, these fires caused the largest emissions of carbon dioxide from Equatorial Asia since the El Niño fires of 1997; short-term exposure to this pollution is estimated to have caused 11,880 extra deaths.

Up until the early 1990s, crop burning was the main source of PAH emissions in the UK and Europe, too. As a child growing up in rural England, I remember every autumn seeing fields black with smoke as farm workers would walk slowly behind the flames, wafting them across the burning earth. As a child, it was an exciting sight. But Ally has some data to show me what those days were really like. He opens up his laptop to dig out a lecture slide showing a graph of benzo(*a*)pyrene emissions in the UK from 1990 to 2009. In 1990, a total of 60,000kg benzo(*a*)pyrene emissions were recorded, with agricultural crop burning accounting for almost half, around 27,000kg. In 1993, when crop burning was banned in the UK, the agricultural contribution literally disappears immediately, and total benzo(*a*)pyrene emissions are half that of 1990, now coming only from industrial processes. Then, when industrial emissions standards were tightened in 1995, total annual emissions go down again to 10,000kg, just a sixth of the levels from five years before. By 2009, there was only one major contributor – domestic wood-burning stoves – at around 3,000kg a year, but growing every year since 2002.[2]

PAH concentrations are often far higher in industrialised and urban areas of developing countries. Studies between 2002 and 2009 found PAH levels of 8–29 nanograms per cubic metre (ng/m^3) in Algiers, Algeria, 38–53ng/m^3 in Ho Chi Minh City, Vietnam and 3.1–48ng/ m^3 in Kuala Lumpur, Malaysia. The UK government's Expert Panel on Air Quality Standards (EPAQS) recommends an annual average air quality standard for PAH of just 0.25ng/m^3, citing links to lung cancer, skin cancers and bladder cancer. Health effects from chronic or long-term exposure to PAHs may also include decreased immune function, cataracts, kidney and liver damage, jaundice and lung function abnormalities including a high incidence of lung cancer risk in developing countries.

The notable others

Sulphur dioxide

If I was writing this book in the 1980s – or any industrial decade before that – sulphur dioxide (SO_2) would have been a leading actor. It is, and was, the central ingredient of acid rain, formed whenever too much fossil fuel containing sulphur is burned. Coal and crude oil both contain a high percentage of sulphur. By the middle of the industrial revolution in 1850, the sources of sulphur dioxide were broadly split half and half between the industrial burning of fossil fuel and global volcanic activity. A century later, an estimated 90 per cent of sulphur dioxide in the air came from man-made sources.

The reason why we don't hear as much about acid rain any more, however, is that – after fish in lakes started to die and the façades of ancient buildings began melting like wax – the global community went to great lengths to take sulphur out of fuel. The UN's 1985 Helsinki Protocol on the Reduction of Sulphur Emissions was very effective. In the US between 1980 and 2013, annual average sulphur dioxide concentrations decreased by 87 per cent, while in Europe, sulphur dioxide emissions decreased by 76 per cent between 1990 and 2009. However, removing sulphur from fuel is expensive, meaning that lower-grade fuel including sulphur remains the norm in many developing countries and in international shipping. Between 2000 and 2010, the contribution of Asia to global sulphur dioxide levels increased from about 41 per cent to 52 per cent, while that of North America and Europe (including Russia) declined from 38 per cent to 25 per cent. China emitted 30 per cent, or 29 megatonnes, of the world's total sulphur dioxide in 2010. While in Delhi, 55 per cent of all sulphur dioxide released comes from its two inner-city coal-power plants.

Since 2000, sulphur dioxide sources have been measured by satellite,[*] giving us a comprehensive global 'top 500' table of the largest sulphur dioxide emitters in 2016; it included 297 power plants, 53 smelters, 65 oil and gas industry sources and 76 volcanoes. The smelters in Norilsk, Russia, represented possibly the largest single anthropogenic sulphur dioxide source seen by satellite, with total sulphur dioxide emissions of up to 1.9 megatonnes[†] from a single source (by contrast one of the largest volcanoes, Mount Etna in Italy, emitted an estimated 0.5 to 1.2 megatonnes a year).[3] But if international shipping were a country, its sulphur dioxide emissions would dwarf those of the whole of Russia. The SO2 from shipping was estimated by one 2003 study to weigh in at 12.98 megatonnes (between 10 and 26 Mount Etnas).

Carbon dioxide

In the story of air pollution, greenhouse gases such as carbon dioxide (CO_2) and methane (CH_4) are the first cousins once removed. They come from the same family – the burning of fossil fuels and agriculture – but you're unlikely to fall ill from them (although the health consequences of climate change are, of course, severe). If you deal with the sources of air pollution, however, you tend to deal with the sources of greenhouse gases

---

[*] The first satellite measurements of sulphur dioxide were actually recorded in 1979. However, those measurements, taken by the Voyager 1 satellite, were of the atmosphere of Jupiter's moon Io, not of Earth, meaning that for two decades we had a better understanding of SO_2 on a distant planetary body than we did of our home planet.
[†] One megatonne = one million metric tonnes.

at the same time. In terms of bang for your buck, tackling air pollution gives you a double return of solving local health problems *and* reducing climate catastrophe globally.

Carbon forms a part of all living things within the ultimate recycling system: carbon atoms constantly lend themselves to new uses, whether it's soil, plants, the body of an animal or a gas. CO_2 is one such carbon-based gas, being one carbon atom combined with two oxygen atoms. While other planets such as Venus and Mars have atmospheres dominated by CO_2, Earth's atmosphere has far less because its liquid water and plants respectively absorb and feed on CO_2. When carbon dioxide dissolves in the ocean, it forms carbonic acid.

Thanks to ice-core sampling in Greenland and Antarctica – where each year's snowfall layer forms an ancient record like rings on a tree – we know that in parts per million (ppm), CO_2 has peaked and troughed from 180ppm to 280ppm on a reasonably predictable cycle of every 100,000 years. But in 1950, CO_2 breached the 300ppm mark for the first time in human history – and has continued to rise. September 2016 was a milestone no one wanted to reach. At a time of the year when atmospheric carbon dioxide is usually at its lowest, after a full summer of plants in the northern hemisphere feeding on CO_2, the value in September reached 400ppm and failed to drop. To picture what 400ppm looks like, the Carbon Visuals blog suggests that in a small office meeting room, 400 parts per million of air would be similar to two large water coolers in the room. A 100ppm increase should take around 7,000 years. This one took 66 years. The CO_2 in the atmosphere increased by 20ppm in just a single decade between 2005 and 2015.*

---

* The last time there was this much CO_2 in the air was around 3 million years ago, during the Pliocene. Back then, CO_2 levels remained at around 365–410ppm for thousands of years, during which Arctic temperatures were 11–16°C warmer than 2011 levels, and sea levels were around 25 metres higher. According to Nasa, if we continue at a business-as-usual rate, and humanity exhausts the reserves of fossil fuels over the next few centuries, CO_2 will continue to rise to 1,500ppm.

With levels of CO_2 in the atmosphere increasing, oceanic life is starting to suffer due to higher acidity, causing problems such as reduced shell growth and bleaching coral reefs.

While volcanoes are a natural source of CO_2, this source is dwarfed by human combustion sources. According to the Nasa Earth Observatory, volcanoes emit between 130 and 380 million metric tonnes of carbon dioxide per year, whereas humans emit about 30 billion, 100–300 times more than volcanoes, by the burning of fossil fuels. Power plants are the largest stationary source of greenhouse gases in the United States. Energy production in the US accounted for 86 per cent of total 2009 greenhouse gas emissions. The rainforests that we need to absorb CO_2 from the atmosphere are also being cleared and burned, producing a lose/lose scenario whereby there are fewer trees to absorb CO_2, and burning them releases more carbon back into the air (plus NOx and polycyclic aromatic hydrocarbons). In Indonesia, forest fires on an almost incomprehensible scale have raged for much of the past decade, many simply to make way for palm oil plantations; 80 per cent of its rainforest has been lost this way. The fires of 1997–8, which burned across an estimated 9.7–11.7 million hectares (roughly the size of the state of Ohio) on Borneo and Sumatra, destroyed 4.5–6 million hectares of tropical lowland rainforest and 1.5–2.1 million of peat soils. The estimated carbon emissions were equivalent to 13–40 per cent of annual *global* fossil fuel emissions that year. The forest fires of 2015 were worse still.

Carbon monoxide

When carbon is burned, it binds with oxygen. If it buddies up with two atoms of oxygen, then carbon dioxide is formed; if it joins with just one oxygen atom, then it becomes carbon monoxide (CO). As Devra Davis puts it, both can kill you in high enough quantities, but 'it takes a lot less carbon monoxide to do the job'. At a natural background level of around 0.2 parts per million (ppm), carbon monoxide should be insignificant to the point of being harmless. But with high levels coming from vehicle exhausts, industrial production such as steel foundries, and also tobacco smoke, it is one of the deadliest gases in the air. Carbon monoxide is one of those cameo actors that can be a real scene-stealer. Most of the time it doesn't appear on screen, but when it does, boy do we know about it. By reducing the amount of oxygen carried by haemoglobin around the body in red blood cells, it causes death at just 250 parts per million (ppm). A small but significant number of people die every year on campsites after bringing disposable BBQs inside their tents at night to keep warm – zipping up their tent literally seals their fate from CO. Vehicle emissions in the 1970s and 1980s brought carbon monoxide in outdoor air to dangerous levels, before catalytic convertors were fitted and carbon monoxide retreated: measurements from central London's Marylebone Road show a sharp and sustained 12 per cent per year drop in carbon monoxide between 1998 and 2009. However, it hasn't gone away altogether. As with sulphur dioxide, its decline in Europe and North America has been countered by a concurrent rise in the southern hemisphere and Asia. India shows a steady increase in carbon monoxide from 1990 to 2010, while Chinese emissions sped up significantly post-2000, largely due to huge increases in car ownership, diesel fuel and industry such as metals and chemical manufacturing.

Methane

The carbon gas that doesn't include oxygen is methane (CH_4) – one atom of carbon and four of hydrogen. It comes from the decay or digestion of organic matter, either in the guts of animals or by bacteria. As such, the sources of methane mostly come from waste dumps, agriculture and fossil fuel. Given that crude oil is simply decayed organic matter, the gas that sits on top of it is largely methane, and this is the gas that's collected and piped into your gas cooker. The direct health effects of breathing it in are minimal (the highly toxic coal gas supplied to households before the switch to natural gas in the 1960s and 1970s in the UK was a different matter), but it is highly explosive – if your house fills up with it, breathing in the concentration of gas wouldn't kill you, but turning on the light switch would. In terms of its greenhouse effect, every molecule of methane is equivalent to around 30 molecules of CO_2, so increased amounts of it do matter. Its concentration in the air is now more than double pre-industrial levels, mainly due to the increase in emissions from cattle farming, rice paddies, landfill sites and even leaks from the natural gas network. In 1750, the global average of methane was an estimated 772ppb; by 2011, it was up to 1,803ppb. Methane concentrations in Delhi between 2008 and 2009 were found to fluctuate from 652ppb to 5,356ppb in residential areas, and as high as 15,220ppb in so-called 'unauthorized residential' areas or slums.[4] The use of CNG (compressed natural gas) fuel used in three-wheelers and buses contributed nearly 50 per cent of methane emissions from 2006 to 2010.[5]

But agriculture is easily the largest methane emitter, accounting for 54 per cent of methane emissions in the EU in 2015. Digestion emissions from livestock alone account for 22 per cent of all US methane emissions. About 10 to 12 per cent of the total US methane emissions come purely from beef production; an individual cow can release up to 500 litres of methane a day. Worryingly, these emissions are likely to rise as global demand for meat increases: the FAO predicts global meat production will be 16 per cent higher in 2025 than it was in 2015, and world milk production is expected to increase by 23 per cent over the same period. The natural background methane emissions from wetlands and volcanoes, according to the US Geological Survey, generate about 200 million tonnes of methane annually; meanwhile automotive and industrial activities cause some 24 billion tonnes a year.

The walk-on parts

While all the gases mentioned so far – both the leading actors and the notable extras – have natural as well as anthropogenic sources, there are some that are entirely man-made. It's tempting to jump to tabloid cliché and call them Frankenstein pollutants, but I won't (although I just did).

Synthetic halocarbons are made by taking a hydrocarbon, removing the hydrogen and replacing it with one or more halogen atoms (chlorine, fluorine, bromine or iodine). The best known are CFCs, and they are built to last. Some have atmospheric lifetimes of not just years, but centuries. The almost immortal perfluoromethane has a lifespan of 50,000 years (compare that to the fleeting one-second life of the free radical OH). Prior to 1930, the preferred refrigerant gases for household fridge-freezers were propane, ammonia and sulphur dioxide – volatile gases which unsurprisingly led to a number of explosions and deaths.* In 1928 General Motors tasked its feted engineer Thomas Midgley (remember his name, we'll meet him again later) to come up with a solution, and he suggested non-volatile, inert CFCs, which had first been synthesised in the nineteenth century but hadn't yet found a commercial use. Midgley demonstrated their unparalleled low flammability by inhaling some of his new gas and then exhaling onto a lighted candle, which was extinguished rather than exploding. CFCs were soon the norm in fridges and air conditioners, and their chemically unreactive, non-toxic properties also lent themselves to fire extinguishers, propellants in aerosol sprays, solvents in electronics manufacture and foaming agents in plastics. The very inertness of CFCs, however, would later have catastrophic consequences. 'Because they live for so long they get taken up into the stratosphere,' Ally explains. 'At ground level the strength of the sun isn't strong enough to break the bonds in this type of halogen-containing molecule. The only place in the atmosphere which breaks them down is when you get high up into the stratosphere where the sunlight is stronger. So, they drift around the atmosphere for years and in the end

---

* Refrigerators work by trapping a volatile gas in a sealed system of pipes – the pipes at the back of the fridge keep the gas compressed into a liquid and relatively warm, but just before entering the pipes inside the fridge or freezer it expands back into a gas and suddenly cools. It's a similar principle to an aerosol spray suddenly cooling as it hits the air, only in this case, trapped in a continual cycle.

a small amount of them find their way to the stratosphere and there the sunlight is strong enough to actually break the bonds.' Here the chlorine finally breaks free and becomes a free radical, desperate for an oxygen atom to partner with. As ozone has three oxygen atoms, it is hungrily destroyed by the chlorine released by CFCs. And as most CFCs have a lifetime from tens to hundreds of years, ones released when you were a child are probably still patiently making their way up to the upper atmosphere.

CFCs are just the most well-known of the Frankenstein pollutants (sorry, the name's stuck now). According to the *Lancet* Commission on Pollution and Health (2018), more than 140,000 new chemicals and pesticides have been synthesised since 1950. Of these, the 5,000 that are produced in greatest volume 'have become widely dispersed in the environment and are responsible for nearly universal human exposure' while fewer than half 'have undergone any testing for safety or toxicity'. Rigorous pre-market evaluation of new chemicals has only become mandatory, according to the Commission, 'in the past decade [2007–17] and in only a few high-income countries'.[6] The UK Chief Medical Officer's Report agrees that, 'we only monitor a handful of the thousands of chemical, physical and biological pollutants that an individual will be exposed to over their life time.'

The sky's the limit

Meeting Ally Lewis helped me to picture the atmosphere as a low-burning flame – the more chemicals we throw onto the fire, the more the atmosphere struggles to filter them out. But this process also depends on the height of the air: yes, the air we breathe has a height. And it's probably lower than you think.

The lowest portion of the atmosphere, nearest to ground level, is known as the troposphere, and the bit of the troposphere that we live and breathe in – and where all these gases are emitted – is known as the 'atmospheric boundary layer'. The height of it is constantly changing, from just a few metres above ground up to a couple of kilometres. I went to Paul Agnew, who heads the air quality team at the UK Met Office, to explain this one for me. Paul started his professional life working for the Atomic Energy Authority in New Mexico, helping the Russians and Americans on a project to use nuclear power to generate electricity in space. When it became clear that the nuclear space race was never going to happen he switched to meteorology instead. He soon learned, however, that atmospheric chemistry is much more complicated than rocket science. The boundary layer, he explains, is 'that portion of the atmosphere close to the surface of the earth which is strongly influenced by the surface and features of the surface such as hills and convection where you've got a lot of turbulence'. Trying to visualise it, I ask if the upper limit of the layer is the flat underside of clouds? 'That is often a good approximation to the height of the boundary layer, yes. There will be occasions when that's not true, but ... the base of the cloud layer is a reasonable guide to the height of the boundary layer.'

Memorably in Paris, Amélie Fritz, environmental biologist at Airparif, described it as being like 'running your car in the garage with the doors closed. The top of the boundary layer sits on top of us like a lid.' In December 2016 in Paris, 'this layer was around 50 metres, sometimes even lower', she tells me. 'We were mixing our pollution within, like, 20 metres high. Which is nothing. That's why pollution levels just went crazy ... We were breathing our own pollution, basically.' Dr Sean Beevers, Senior Lecturer in Air Quality Modelling at King's College London, also told me, 'Because the sun drives a lot of the boundary layer turbulence, in summer – even in the UK, where summers are not particularly warm – you get quick, rapid heating of the ground and then it heats the air above it and that creates the turbulence and the boundary layer height lifts. The counter is true as well, so overnight you can get some quite rapid cooling and then you get a low boundary layer, especially during winter.' In the Gobi Desert, the boundary layer has been found to stretch up to four kilometres from the ground in daytime but plummet down to near zero at night – in which case, if you stood up, your head would be poking into the calm, non-turbulent 'free atmosphere' level of the troposphere.

The atmospheric boundary layer is sometimes called the 'mixing layer', because it is the only layer of the atmosphere in constant contact with the surface of the earth, is a swirling mix of temperatures and wind speeds, where everything is whisked together. But the level above it, with no surface to react with, is relatively calm. The lid between the two layers is surprisingly effective, with most pollutants staying within the lower boundary layer owing to the simple fact that they are themselves constantly reacting with other chemicals and are short-lived. That's what makes CFCs so unusual because they can survive for decades, slowly making the journey up to the upper atmosphere.

Within an urban environment, the boundary layer height is crucial to air quality, says Ally: 'Most cities have pretty much the same emissions day in and day out, so it is the meteorological factors that then control how high or low pollution concentrations end up … Shallow night-time boundary layers are a big issue in places like Beijing that have relatively low windspeed year-round. London can also have a low boundary layer, but it is generally a windier place, so this is a less common phenomenon.' Air pressure plays an important role in this: during high-pressure systems, the air is usually still, which allows pollution levels to build up, while during low-pressure systems the weather is often wet and windy, causing pollutants to be dispersed or washed out of the atmosphere by rain.

Most pollutants, other than ozone, that cause chronic health conditions over long periods of time, are higher in winter because of the lower boundary layer. Unfortunately, winter is also when our boilers and log burners all fire up, and we may prefer the warmth of our car over the chill of the pavement, 'so it is the kind of worst of all worlds', says Dr Beevers. Cities do have one small advantage, in that they tend to trap and radiate heat better than rural areas, leading to a higher boundary layer – something known as the 'urban heat island' effect. But that is outweighed by the same cities being the source of pollution in the first place. Adding a few tens of metres on the boundary layer height doesn't mean that much when millions of cars and heating systems are emitting pollutants into the air. Not only that, but city buildings also create a 'street canyon' effect. With tall buildings either side of a street, and a boundary layer sitting on top or even below the height of the buildings, pollutants are trapped on all four sides, while we walk, cycle, drive or push prams through it. Oxford Street, one of London's busiest shopping streets and tourist attractions, is also one of the city's worst street canyons for air pollution.

Measuring the exact quantities of pollution, and our exposure to them, is therefore an extremely complex science. Where I now live in Oxfordshire, in the south of England, air quality is measured by local government authorities using diffusion tubes. Diffusion tubes, widely used in the UK and globally since the 1970s to measure ambient concentrations of NO_2, are small plastic test tubes containing a chemical reagent called triethanolamine, which absorbs NO_2 directly from the air. When the triethanolamine is removed in a laboratory, analysis can show the average concentration of NO_2 that was present in the air for the time the tube was exposed. Typically retrieved and tested once a month, they can only record monthly averages, with no indication of peak hours or even peak days. Diffusion tubes could be described as an educated guess at best. They are also known to have a relatively high uncertainty, as much as $\pm$ 25 per cent according to the UK government's own advice. But they are the number one choice of air quality monitor used by local authorities across the UK. To record with more accuracy and detail, some larger towns and cities also have a stationary 'automatic monitor', capable of monitoring at least NOx and particulate matter in close to real time. The nearest one to my home is on Oxford High Street – some 30 miles away from my house, and not much use if I want to check what the air is like before embarking on the walk to my daughter's school.

Dame Sally Davies, the UK's Chief Medical Officer, chose to focus her 2018 Annual Report (all 348 pages of it) on pollution. She warned in her introduction, 'We do not have the systems in place to effectively monitor, understand, and act on data about the health impacts of pollution.'[7] The company with the contract to fit and service the UK government's entire automatic monitoring system (AURN, Automatic Urbal and Rural Network), is Air Monitors. Jim Mills, the company's founder and MD, actually agrees that the current method of monitoring in the UK – and in many other countries – just isn't working. 'The UK must increase its monitoring capacity because there are so many places where we're relying on 10-quid-a-throw diffusing tubes to measure NO_2. All that does is tell you whether over the entire month you were over or under the European directive, it offers no value … You can't improve air quality with NO_2 diffusion tubes … it's so frustrating when you see councils being so casual about something so important.'

The reality is that most of us don't know what pollutants we are exposed to on a daily basis. Even if you happened to spend all of your time standing next to a diffusion tube, you still wouldn't. And if you are lucky enough to be in a city with one of Jim's sophisticated automatic monitors, and happened to be on the same street, you *still* might not know. Jim tells me that the section of Marylebone Road in London where Madame Tussaud's is located is 'one of the most polluted streets in the capital. The [AURN] monitoring station we have there uses the most sophisticated, latest, best equipment – it's about 8 x 4 metres, it's right outside Westminster Council House, and that monitoring station is regarded as the sort of crème de la crème. However, it's monitoring on *that* side of the street … the exposure of a person who is standing outside Tussaud's [on the other side] waiting in the queue could be three times higher than the station is reporting, and vice versa depending on the weather. It's an incredibly complex dynamic situation.'

In terms of knowing what the desired level should be (i.e. what poses an acceptably low risk to our health), the WHO offers guideline recommendations, which the international community treat with varying degrees of deference. WHO standards are typically far tougher than those imposed by almost any country or region, sulphur dioxide standards being the most ignored: whereas the WHO recommends public exposure of no more than 8ppb of sulphur dioxide in a 24-hour period, the EU sets its cap at 48ppb, Canada at 115ppb and the US at 140ppb. In May 2008, under a court order, the United States Environmental Protection Agency (EPA) grudgingly lowered its ozone standard from 80ppb to 75ppb, despite the Agency's own scientists and advisory board having recommended lowering the standard to 60ppb; the WHO recommendation is 50ppb. The fact that the US ozone standard remains one of the best in the world shows how dangerous a mismatch there is between the scientific understanding of air pollutants and the regulations put in place to protect us from them.

And even when they do cram a city such as London full of sensors, the same conflicts of interest still arise. Prior to the London Olympics, Simon Birkett of campaign group Clean Air in London personally witnessed – and flagged down a taxi in order to follow – a lorry spraying a pollution suppressant (calcium magnesium acetate) on the roads. The lorry took the exact roads, and only those roads, where London's pollution monitoring devices are located. 'It had no effect at all on London's pollution levels – all it did was hide the results and the legal breaches,' says Birkett.

CHAPTER THREE

Particulate Matters

Despite the ill effects of all the gases covered in the previous chapter, there is one category of air pollutant that stands head and shoulders above them all. And it's not a gas. It's a solid: particulate matter (PM). These are the tiny particles that float in the air, from road dust to soot, and cause the most damage to our health. We first met these in the Prologue, so as a reminder: scientists define PM not by what they are made of (such as coal smoke, agricultural dust, engine fumes) but by size. The largest category, PM10, is any particle measuring 10 micrometres in diameter or below (roughly a tenth of the width of a human hair). Smaller PM2.5s are less than 2.5 micrometres in diameter (a fortieth of the width of a human hair). And their little cousins the nanoparticles are below 0.1 micrometres (by which point the human hair comparison becomes a bit pointless[*]). And broadly speaking, the smaller they are, the more effectively they destroy our health.

I had been warned against buying a cheap PM2.5 monitor over the internet (admittedly, mostly by the makers of the expensive ones). Yet here, on my desk, was a package fresh from China containing just that: the 'Laser Egg 2' produced by the Beijing-based start-up Kaiterra. The reason why I decided to give this one a go was a recommendation by email from Frank Kelly at King's College London – his colleague had tested a few against the expensive, finely tuned monitors in the lab, and found the Laser Egg 2 matched up rather well. It could fit in my hand luggage, and measures PM2.5 in

---

[*] If you still want to know, it's 1/1000 the width of a human hair. Although nanoparticles go right down to 0.001 micrometres, which would be 1/100,000 of the width of a human hair … but now I'm just splitting hairs.

micrograms (one-millionth of a gram) per cubic metre ($\mu g/m^3$). During my travels for this book I could turn on my 'Egg' wherever I went, as a quick 'pulse check' PM2.5$\mu g/m^3$ reading rather than anything scientifically rigorous. As we saw in Chapter 1, it soon became my valued travelling companion. But when it first arrived, unblemished, on my desk in my home office, I wasn't sure if it was working. It resolutely stayed at $1\mu g/m^3$. As it emits a low hum as it sucks air inside, I knew it was on. A brief flicker up to $3\mu g/m^3$ got me excited, but then it returned to $1\mu g/m^3$. So at lunchtime I took the Egg into the kitchen and purposely cooked up something that I regularly eat but know to be smoky: pan-fried tortilla wraps. My pan of choice is a heavy cast-iron one, the type you're not supposed to wash but just wipe down after each use – the theory being the oils from the cooking eventually create a kind of natural 'non-stick' surface. But it does mean that some old burnt bits do burn again on re-use.[*] My Egg initially appeared little bothered by this, but as the wraps crisped up it shot up to $220\mu g/m^3$, then $280\mu g/m^3$, and finally peaked at $401\mu g/m^3$. While I – and my then-pregnant wife – sat at our kitchen table to eat, the reading remained at a stubborn $180\mu g/m^3$, even with the windows wide open. The next time I had tortilla wraps I baked them in the oven, and the Egg barely reached $7\mu g/m^3$.

The ghoulish cast of gases in the air pale into insignificance compared to PM2.5. When the European Environment Agency recorded the premature deaths attributable to air pollution in 40 European countries in 2012, 432,000 were

---

[*] Before you question my hygiene, this method is standard in many parts of the world. A friend once cycled through Mongolia and was invited in by a local family to enjoy a traditional meal. With no shared language, he thought he'd express his gratitude by doing the washing-up. The blackened wok was a daunting task, but he scoured it with gusto, and eventually got it looking clean and metallic once more. The look of horror on the family's face, when he smilingly presented his work, still haunts him – a decade of seasoning was undone.

credited to PM2.5. The next two biggest killers, nitrogen dioxide (NO_2) and ozone (O_3), only managed a paltry 75,000 and 17,000 deaths respectively. Other global studies have found the same results. Roughly four out of every five deaths caused by air pollution are caused by PM2.5.

David Newby, professor of cardiology at the British Heart Foundation Centre of Research Excellence, began studying air pollution in the early 2000s. I spoke to him a lot for this book, and he is now one of the most prominent names in the field of the health effects of particulate pollution. But he admits that when he first started looking into this some 20 years ago, he presumed that the big chunky particles from black smoke were the worst offenders. 'But actually, those sorts of particles are big and coarse, and get lodged in the upper airways. The particles we are talking about now diffuse right down into the lungs ... Cars generally produce PM2.5 and the majority of these are nanoparticles less than PM0.1 – that's incredibly tiny. In these combustion particles you get metals, organic matter coated in petrol and diesel. It looks much like the tar coming out of smokers' lungs – you see the same images as a result of [PM] pollution.'

I learn more when I go to meet Frank Kelly, professor of environmental health at King's College London and chair of the government advisory body Committee on the Medical Effects of Air Pollutants (COMEAP). On the train to London I open my laptop and put my newly purchased Egg beside it on the table, occasionally glancing at it as I go through my emails. Heading south towards London from Banbury it reads just $5–10\mu g/m^3$ and then goes up to $12–15\mu g/m^3$ almost as soon as we leave High Wycombe. It slowly creeps up and up, even inside this air-conditioned train, as we approach the metropolis. In the tunnel approaching London Marylebone station the train slows down to a halt, presumably waiting for the platform ahead to clear. The smell of diesel reaches my nostrils and the Egg climbs through the 20s, 30s ... to the 80s. We are now – unbeknown to presumably everyone on the train but me – marinating in $83\mu g/m^3$ of diesel particles (classified by the WHO as 'carcinogenic to humans'). Out of the tunnel and moving again now, it drops down into the 60s. When I get onto the London Underground, at platform level the familiar rush of warm air precedes a Tube train approaching. This air, it seems, is also full of PM2.5. The reading rockets up past 100, settling at 121. On the Tube train (as we're in London, where nobody ever talks to each other, I am not asked what the egg-like alarm clock thing is on my lap) the reading stays in the mid-60s throughout the journey, unnervingly high compared to the single-figure and low double-figure readings that I set out with that morning.

When I finally arrive at Frank Kelly's office at King's College London, I thank him for recommending my Egg monitor, and tell him of the PM readings throughout my journey that morning. 'The PM2.5 in the underground is very different to what it was reading in that train tunnel,' he tells me. 'If it was a diesel [overground] train, then the diesel emissions will be lots of tiny particles, PM2.5 and smaller, which are basically formed by incomplete combustion. Whereas what you'll find in the underground normally is dust particles, dirt and also the abrasion from the rails.'

A lot of this research, however, is very new. The Mayor of London asked Kelly's COMEAP Committee to review air pollution levels on the Tube in 2017, and when we meet this is still a long way off being published. By way of explanation, Kelly offers me a potted history of PM research. 'One of the slides I show in my lectures is a timeline from the 1950s to 2010,' he says. 'It shows the concentrations of black carbon and sulphur dioxide falling dramatically from the late 1950s to the early 1980s and 1990s. That was the result of the [UK] Clean Air Act and shutting down power stations in urban environments. If you asked anybody in the late eighties and early nineties about air quality, everybody in the UK including the scientists would say it's not a problem. It was only in the mid-1990s that we started to get information coming in from the United States that in fact there was a new problem, and it was tiny particles. You had to be reading that literature to be aware of it. It probably wasn't until the turn of the twenty-first century that PM really started to become mainstream again.'

I learn that there are broadly three sources of PM: natural aerosols,* secondary PM and primary PM. The natural ones include sea salt, sand, dust, pollen and volcanic ash – basically any natural material that is fine enough to be whisked up by the wind and carried in the air. Having some of these natural particles in the air is necessary for water vapour to collect around and form rain clouds. Aerosols emitted from the sea, literally thrown airborne by breaking waves, are among the most common natural aerosols in the Earth's atmosphere. Another common group is mineral dust, such as sand storms. However, 'natural dust' doesn't always mean the same as 'natural causes'. Agricultural mismanagement and livestock grazing in the early twentieth century caused a 500 per cent increase of dust in the western United States, leading to the 'dust bowl' of the 1930s. Similar desertification is now happening globally, due to food and water shortages.†

For the most part, we have little to fear from natural aerosols (except for viruses and fungal spores, but they're a whole other story). Newby recalls attempting one of his signature 'chamber studies', exposing volunteers to what he thought were low levels of diesel pollution collected from the streets in Edinburgh. On this occasion, none of the usual adverse health effects appeared in the volunteers. When the chemical composition of the PM sample was analysed, it suddenly made sense: 99 per cent sodium and chloride. Due

---

* The scientific term 'aerosol' may bring underarm deodorant sprays to mind, but in fact it's just a synonym for any fine particles of solids or liquids suspended in the air.

† Dust bowls are not exclusive to large nations, but can happen to any country that denudes a natural landscape of trees and hedges. H. F. Wallis records that due to agricultural malpractice in 1968, 'the east of England had its worst dust storms in living memory … the regions where hedges which could have broken the wind were no longer there [due to new intensive farming practices]. Rolling clouds of dust filled the sky …'

to the wind direction on the day the air sample was taken and Edinburgh's proximity to the sea, his volunteers (no doubt much to their relief) were only exposed to sea air. Despite it technically being a high PM2.5 count, the PM2.5 in question was salt, and it did them no harm whatsoever. In fact, Ally tells me, 'one of the reason you don't get goitre[*] in small island nations like the UK is from inhaling sea-salt particles with iodine in it. You need things like sea salt in the air.'

Secondary particles, however, are a little more complex. They are formed by chemical processes within the air itself. Jesse Kroll, chemical engineering professor at MIT, explains that the air is full of new particles being formed spontaneously: gases and VOCs (volatile organic compounds) react with each other and clump together, making the journey from gas particles to liquid droplets and finally to solids. I presumed that most PM2.5 would be primary particles – solid bits of soot and dust – but not necessarily, says Jesse: 'At least in anthropogenic regions ... PM2.5 is mostly secondary. A big one is sulphur dioxide emitted from burning coal or in some cases high-sulphur diesel ... That sulphur oxidises all the way to sulphuric acid and condenses into the particle sulphate. Most sulphate is secondary, the vast majority I'd say.' I wasn't expecting that response. It's one worth bearing in mind whenever anyone talks about filters for trapping particulate matter, for car engines or anywhere else – filters may be able to trap solid particles, but they can do nothing about the ones that are emitted as gases and turn into particles later on. Ammonia gas, for example, reacts with NOx gases in the atmosphere to form solid particles such as ammonium nitrate. Ammonia may be a gas, but it contributes some 10 to 20 per cent of the PM2.5 mass in Europe through secondary PM formation, even in urban areas: the London haze in

---

[*] A swelling of the thyroid gland in your neck, looking a bit like a large Adam's apple, most commonly caused by iodine deficiency and most common in landlocked countries.

March 2014, when my daughter was born, was formed from ammonia released on farms mixing with the NOx from London traffic to form ammonium nitrate particles. Paul Agnew at the UK Met Office tells me, 'When we have a bad air quality episode … it's generally due to a big rise in the formation and transport of secondary PM2.5, and in particular ammonium nitrate … it's a very effective way of producing fine particulates.'

High NOx therefore leads to high PM2.5 because of its role in forming secondary PM; in a sense, high NOx *is* high PM, or at least a lot of NOx *will lead to* a lot of PM. A study by the US Federal Highway Administration suggests that the proportion of all PM2.5 in the US comprised of secondary PM varies anywhere from 30 per cent to 90 per cent. A multi-European study in 2015 of car engine emissions also found that the amount of secondary PM formed was on average three times larger than total primary emissions that initially came out of the exhaust pipe.

Among the easier-to-understand primary pollution particles, emitted directly into the air as a solid, the undisputed heavyweight champ is 'black carbon'. It's soot, basically: the tiny carbon particles formed during the burning of fossil fuels and solid fuels. But there are some things you may not know about soot. It is insoluble in water, and has the ability to absorb light and heat: enough of it in the air can therefore warm up the atmosphere, absorbing and then radiating heat from the sun. As such, when huge amounts of black carbon are produced by forest fires it is a significant contributor to global warming. A 2013 paper from the Chinese Academy of Meteorological Sciences, Beijing, suggested that because black carbon can directly absorb both solar and infrared radiation, it disturbs 'the energy balance in the earth–atmosphere system' and can even 'heat the air of the cloud layer and thus directly cause cloud evaporation and reduction'.[1] This has been linked in Asia to both local warming trends and adverse affects on the seasonal monsoon rains.

A global black carbon inventory in 1996 estimated that annual emissions came from open burning (42 per cent), fossil fuels (38 per cent) and solid fuels (20 per cent). To picture how it's formed, think back to a Bunsen burner in school science lessons, suggests Ally Lewis: 'You had two settings, the blue setting, which is the full combustion setting: you're using no more air than you need to, so it burns really hot, you get no waste product other than water and CO_2. Then you turn it on to the yellow setting and you get the flickering yellow flame and soot ... unstructured webs of carbon that stop being nice discrete individual molecules and become black grit.' They are essentially crisps of carbon, caused by poor combustion. And crispy carbon acts like a sponge for other chemicals, much like its big brother, charcoal. Charcoal has been used for centuries to purify water, because organic chemicals such as VOCs are chemically attracted to it. If you put charcoal into dirty water, the impurities slowly stick to the charcoal and the water becomes drinkable. The same thing happens in the air with black carbon particles, with one obvious downside: we breathe them in, impurities and all. 'Right out of the car exhaust pipe these fragile crystalline structures of pure carbon get very rapidly coated with gunk,' Ally explains. 'Black carbon is like a magnet for chemicals.'

Almost needless to say, black carbon is also highly toxic. Black carbon is what turns the lungs of smokers black. Dr Langley DeWitt, at the Rwanda Climate Observatory, told me that, 'short-term studies near roadways in Nairobi find very high ... black carbon concentrations. Pollution exposure levels along roadways are important as, in East Africa, many people walk along roadways or travel by motorcycle with high exposure to exhaust.'

The European capital of primary particles is Poland. A large and powerful coal industry, combined with the widespread practice of rubbish burning, sees relatively mild summer pollution levels spike to among the world's worst in winter. World Bank figures (2015) show that 81 per cent of the country's energy mix comes from coal, compared to neighbouring Germany's 44 per cent (or France's 2 per cent). Maciej Rys, an entrepreneur living in Krakow, tells me of a particularly bad smog in November 2015: 'One day I was riding a bike in Krakow over the bridge that connects to the new technology district, but it was such a huge smog I couldn't see anything. It wasn't even like smog, it was like milk … it was like milk in the sky … it got really bad after Communism fell because the gas prices got really high and people started burning coal and charcoal and trash to warm their houses. We still have to replace all the furnaces from these old-fashioned, trash-based heaters into gas or district heating.'

Given that black carbon comes from fire, it's unsurprising then that a large – and increasing – source of PM2.5 is domestic log fires and stoves. And the use of wood as fuel for residential heating is actually increasing in developed countries, often incentivised as a 'renewable fuel'. Between 2013 and 2014, a strong PM2.5 emission increase (+29 per cent) was detected in the United Kingdom: the only change within that time was the increased consumption of wood as a domestic fuel. About 40 per cent of PM10 in 2015 came from domestic stoves (double the proportion from diesel cars) with 1.5 million already fitted in Britain and 200,000 being sold annually. This, despite wood smoke still being banned within most UK cities – including London – due to the original 'clean air zones' established by the Clean Air Act of 1956. 'Clean air zones are not policed at all,' says Ally. 'Most people don't even know whether they live in a clean air zone or not. It's particularly a problem in central London now where … PM2.5 and PM10 came down pretty substantially for over 20 years … Now, because of wood-burning stoves, PM is actually going up again in London.' Rather than peaking during the traffic rush hours, 'Friday nights and Saturday nights now are the peaks for PM10 in London, which is insane,' says Ally. He took part in a study to look at the comparative effect of wood-burning stoves compared to vehicles and uncovered some truly shocking findings. 'One wood-burning stove is roughly equal to one 7.5 tonne truck idling outside your house,' he tells me, 'and that's probably a conservative estimate. York is a clean air zone but loads of people here have wood-burning stoves. You couldn't fit enough lorries in the street to actually represent the number of wood burners.'

A headline in *The Times* stating that the environment secretary 'wants clampdown on coal fires and smoky stoves' deserved to be from a previous century. In fact, it was from 30 January 2018. Log stoves were – and are – rapidly becoming a must-have middle-class accessory. Gary Fuller of King's College London has even linked them with the popularity of home and lifestyle TV shows such as *Grand Designs*. In pursuit of the perfect-looking living room, the UK's largest single source of PM2.5 – 37,200 tonnes of it a year – is now domestic wood burning. A *New Scientist* review in 2017 concluded that log-burning stoves 'are harming our health and speeding up global warming'. Yet, despite this, a Defra (Department for Environment, Food and Rural Affairs) consultation on domestic burning (which closed in Feb 2018) stated that it was not 'seeking to prevent' the use or installation of new stoves, but rather to 'encourage consumers to switch to cleaner wood burning'. While there is a gradient of more-to-less smoky fuels, there is no such thing as a clean or 'smokeless' solid fuel. Rob MacKenzie, professor of atmospheric science at the University of Birmingham, told the *Guardian* in 2017 that 'smokeless coal produces more nitrogen oxides than wood fuel, and they both produce the very small particles that are the least noticeable, but the most harmful.'

It's the same story across most of Europe. In Ireland, Professor Colin O'Dowd told the *Irish Times* that extreme air pollution events are driven by the burning of residential solid fuel, which he said accounts for 4 per cent of the fuel burned but 70 per cent of the pollution: 'The major concern is that these fuels are promoted as being "green" or "low-carbon" in terms of greenhouse gas emissions, but are devastating in terms of air pollution,' he said, adding that it is 'turning back the clock in terms of clean air policy'. In Helsinki, Finland, approximately 90 per cent of family houses have fireplaces. Approximately a quarter of all combustion-based fine particle emissions in the Helsinki metropolitan area come from fireplaces and wood-burning saunas.

North America hasn't been immune to this wood-stove epidemic either. In Maine, the use of wood as a primary heat source increased from 7 per cent of households in 2005 to 13 per cent by 2015, while in rural Montana residential wood smoke contributes up to 80 per cent of winter PM2.5. In the San Joaquin Valley Basin, California, combustion of solid fuels including wood has been identified as the largest individual source of PM during winter. This is happening in major cities too. In San Jose, Atlanta, Montreal and Seattle, wood burning contributes between 10 per cent and 39 per cent of winter PM2.5. Liam Bates, the CEO of Kaiterra, the makers of the Laser Egg, recalls a friend who attended an air quality conference in Aspen, Colorado: 'They were sitting in this large ski cabin with a log fire, and he brought his Laser Egg along ... he turned it on and it was something like $450[\mu g/m^3]$ inside the lounge, just from fireplaces. He was shocked, as was everybody there – which is ironic, because they were all indoor air quality specialists. People thought it was broken, so he stuck it out the window, and it dropped down to $2[\mu g/m^3]$.'

Another form of primary PM is tiny shards of metal. Lead, the elder statesman of air pollutants, smelted by the Romans, surged back to prominence in the mid-to-late twentieth century thanks to the use of leaded petrol in automobiles. So-called tetraethyl lead was first added to gasoline to avoid 'engine knock' in the lab of General Motors engineer Thomas Midgley, in 1921. Yes, the very same Thomas 'CFCs' Midgley. In life, he was awarded the Priestly Medal, the American Chemical Society's most distinguished honour; in death, he was described by the science writer Gabrielle Walker as being 'responsible for more damage to Earth's atmosphere than any other single organism that has ever lived'.* The first health

---

* Midgely contracted polio late in life, and his final invention was a harness system that allowed him to move between his bed and his wheelchair. The harness caused his death by strangulation in 1944 at the age of 55.

warnings appeared almost immediately in 1923, when workers at the DuPont tetraethyl factory began dying following 'violent bursts of insanity'. Not wanting to admit defeat, Midgley called a press conference where he poured tetraethyl on his hands and inhaled the fumes for 60 seconds (a trick he would go on to repeat with CFCs), claiming he could do so every day without any ill effect. The press left satisfied; Midgley secretly took a leave of absence from work, diagnosed with lead poisoning.

Lead, writes the epidemiologist Devra Davis, has the same electronic charge as calcium, and therefore competes with calcium throughout the body: 'In the bones, the brain, and the blood and throughout the nervous system all of which depend on calcium, lead can trigger irreparable damage.'[2] As a means of getting it into the bloodstream, she says, you couldn't invent a better way than miniaturising it, putting it into a combustible liquid and releasing its fumes into the air. Ethanol could just as easily have been added to stop 'engine knock', but ethanol couldn't be patented; tetraethyl was a new formula, and could be patented, and could therefore make more money. The whole world was literally poisoned for profit.

A 1983 report by the UK Royal Commission on Environmental Pollution concluded that lead was so widely dispersed during the twentieth century that 'it is doubtful whether any part of the earth's surface or any form of life remains uncontaminated by anthropogenic lead.' Leaded petrol would be commonly used throughout the world until the late 1990s, when most major countries began to phase out its use. But lead particles can still be found in the air, most now emitted by the iron and steel industry, which is also the leading source of mercury emissions. And unlike many other pollutants or even radioactive materials, lead does not break down over time. Lead particles can also be churned up from road surfaces: lead chromate is often used as the yellow pigment in road markings, while the lead from seven decades of leaded petrol use can still be found, and re-suspended in the atmosphere, worldwide. In Beijing, lead levels increased from 2001 to 2006, despite leaded petrol having been banned in 2000 – the ban was offset by lead emissions from coal and oil combustion, the steel industry and cement dust.

All these nuances and different substances get rather lost in the catch-all term 'PM2.5'. 'They do, don't they?' agrees Ally, relieved that I've finally worked it out. 'PM2.5 is just a convenient way of measuring. It's a measurement of how much mass of particles is in the air, it tells you nothing about what the particle is, it doesn't tell you whether it was from a forest fire, whether it's black carbon, whether it's a liquid aerosol, it tells you nothing at all. That metric was adopted in the 1970s because that's all people knew … it seems a very poor measure if you're working in this field now.' Peter Brimblecombe, who is now a professor of environmental chemistry at City University of Hong Kong, worked on scientific advisory committees to the UK government in the 1970s, and doesn't disagree with Ally's assessment: 'I even remember in the 70s people saying, "well wait though, the particles are coated with polycyclic aromatic hydrocarbons, they're a complex group of organic material …" I think that was discussed. But I think it was also an instrumentation problem, there wasn't much measurement equipment around to determine the differences.'

Working out the exact chemical composition of PM remains difficult given that a cloud of dust can come from multiple local and transboundary sources. One way to separate them out is to use techniques such as 'Ion beam analysis', where a charged particle moving at high speed strikes a material and slows down or deviates in a way that is characteristic of the element it collides with. When that technique was used in Manila, the Philippines, in 2004 to identify the PM2.5 in local air pollution, it found that the PM2.5 came from biomass burning (39 per cent), oil burning (21 per cent), salt (17 per cent), black carbon (14 per cent), soil (8 per cent) and two-stroke engine emissions (1 per cent).

Each city therefore has its own unique PM fingerprint. In Stockholm, Sweden, the first days of spring are heralded when up to 74 per cent of PM10 comes from road dust as winter studded tyres rip up the road surface exposed by melting snow.

In Delhi in 2012, the biggest single source of PM2.5 was transport (17 per cent), followed by power plants (16 per cent), brick kilns (15 per cent), industry (14 per cent), households (12 per cent), waste burning (8 per cent), diesel generators (6 per cent), road dust (6 per cent) and construction (5 per cent). In Nairobi, Kenya, in 2014, mineral dust and traffic were responsible for approximately 74 per cent of PM2.5 mass, with industrial activity, combustion and biomass burning the other major sources. The most stomach-turning example is Mexico City in the late 1990s, when the inner city had more than two million stray dogs depositing 353 tonnes of dog poo a day: the PM10 in the air was found to include 'dog dust' – particles of dried faecal matter. As the *LA Times* nauseously put it in 1999, 'The dog dust and other particulates settle on the tortillas, tamales and salsa being served up to customers at the open-air stands, feeding chronic intestinal miseries.'

The exact same PM2.5 mass measurement in two different cities can therefore pose very different threats to the populations breathing them in. When Chinese journalist Chai Jing took a PM particle filter sampler with her during an ordinary working day in Beijing in 2012, it registered $305.91\mu g/m^3$. The filter turned from pure white to pure black. To find out what the stuff was, she invited Dr Xinghua Qiu from Peking University to chemically analyse her sample. It contained 15 carcinogens, including biphenyl, acenaphthene, benzo(*e*)pyrene (one of the most potent known carcinogens) and fluorene (one of the key pollutants during the 1948 Donora disaster).

To complicate matters further, while some PM is emitted locally, other particles can travel for miles, sometimes hundreds or thousands of miles. In many cities the background level of PM2.5 is mostly made up of these 'transboundary pollutants'. At the Rwanda Climate Observatory, Dr DeWitt estimates that haze from regional agricultural burning is the dominant source of PM2.5 in Kigali, with local sources only accounting for roughly '30–40 per cent of total pollution averaged over the whole year'. Wildfire events in the western US are also estimated to contribute about 18 per cent of

America's total PM2.5 emissions and are increasing. California's devastating Thomas Fire became the largest in the state's history in December 2017. Jeffrey Pierce, a professor of atmospheric science at Colorado State University, told a press conference this was 'the new norm for California', and that wildfires are on the verge of becoming the largest annual source of PM across the entire US. Less than a year later, it was usurped by the Mendocino Complex fire, which burned to almost twice the size of Thomas, in August 2018.

Even in London, it transpires that up to 75 per cent of PM can be transboundary, coming in from elsewhere. When I discovered that, I got a little dejected. It spoiled the picture that had begun to build: that most pollution is local and can be dealt with through local action. Transboundary pollution was beginning to sound like climate change – that the actions of one city or even one country can ultimately be meaningless if others do nothing or emit even more. Did that 75 per cent figure suggest that, even if we got rid of all cars, buses, log stoves and BBQ grill restaurants, such action could only affect 25 per cent of the PM2.5 pollution?! I put that question first to Amélie Fritz in Paris: 'Well, it depends if we are looking at an annual average or a peak. Usually a peak, or pollution episode, is because we are stuck in a meteorological condition of low boundary layer, no wind … and we are basically polluting ourselves. That *is* mostly local. But on a daily basis, on an annual level, we have wind and yes we are receiving pollution from our neighbour … Sometimes even the wind can circle around and bring our own pollution back to us, because we are a big emitter.'

Dr Sean Beevers, who works on the London Atmospheric Emissions Inventory (LAEI) at King's College, helped explain it further: 'If you're just in an average London location then a large proportion of the PM does come from outside. But if you're stood close to a major road then that's not necessarily the case, you have more of the local stuff … NOx and NO_2 come from urban areas, it's generated locally.' Yvonne Brown, an air quality policy analyst for Transport for London, who also works on the LAEI, similarly tells me: 'The proportion of transboundary pollution, often called regional background, will vary depending on location as well as pollutant … For example, at roadside locations where our concentrations tend to be highest, the dominant contributor to pollution is road transport. For NOx, this road transport element tends to be around 75–80 per cent, with other sources such as heating and regional background being a much smaller fraction. But as you move away from the roadside, pollution levels reduce but the contributions of sources begins to vary,' meaning that as you get away from the road the PM2.5 readings go down, say from $40\mu g/m^3$ to $20\mu g/m^3$, but the percentage of PM2.5, within that $20\mu g/m^3$ that is transboundary goes up. I breathed a sigh of relief after the days of research and interviews that cleared that up.

Then in October 2017, the white walls of my home office glowed an eerie orange. My neighbour called over the garden wall, 'Come and have a look at the sun'. It was high in the sky, yet red, like a setting sun. It also happened to be the day that I spoke to Paul Agnew at the Met Office, so I asked him what was going on. While Saharan dust is often used as a scapegoat for high pollution episodes, this time it actually was Saharan dust: 'In order for pollution to travel from North Africa to us it needs to get lifted quite high into the atmosphere, several kilometres,' explains Paul. This cloud of orange sand was floating across much of the British Isles, pushed along by the recent Hurricane Ophelia. But 'most of its journey from North Africa will be taking place above the boundary level,' said Paul. The Saharan dust, he said, would remain kilometres above our heads until reaching northern Norway where it would 'most likely disperse in the upper atmosphere'. Transboundary pollution, then, often happens way above our heads without affecting the air we breathe. It's the stuff happening at ground level that we need to worry about.

David Newby, the cardiologist, explains further: 'When you talk about ambient particle exposure, there are two things you need to think about: one is the overall background levels, and that is going to be dominated by meteorological factors, which way the wind's blowing and where it is blowing from … and yes it can be transported from miles away. But when we're talking about city traffic, the particles are generated on the road, and they disperse quite quickly. It's an exponential decay, usually from the roadside. Ten metres away the levels will be very much lower than if you are actually standing by the cars.'

When dealing with national percentages, the contribution of road transport to NO_2 and PM2.5 can seem relatively modest, often around the 20 per cent mark. Other sectors such as industry and power generation can appear at parity, sometimes even greater. But in terms of where we actually live, work and travel, we spend far more time on or near roads than we do standing next to power stations. Within towns and cities, our proximity to vehicles exposes us to the smallest particles that can enter into our bloodstream. Particulate number (PN) – the total number of particles in any given gulp of air, rather than the mass or weight of particles – is therefore the final piece of the particulate jigsaw. Unlike the background sources of urban air pollution, vehicle emissions occur at ground level and near to us humans breathing it all in. The very smallest of these particles, PM0.1 and below, known as 'ultrafines' or nanoparticles, are increasingly being linked to the worst health effects. They are too small to weigh in any meaningful sense. By number, not by mass, they form the greatest percentage of overall PM2.5. And it is the number of them that enter our lungs, not the mass, that is of most concern for human health.

'Size fraction may have specific or enhanced toxicity,' writes Frank Kelly, because 'smaller particles have a much greater surface area' for toxic chemicals to stick to. What does this mean, that smaller particles have a greater surface area than large ones? It took me a worryingly long time to get my head around this one, too. What it doesn't mean, is that a single small particle literally has more surface area than a single large one – the large one is, well, larger. But if you fit lots of small particles into the space of one large particle, then there's lots more surface area in total. I turned to sports – as I often do – to visualise it. A standard football (that's a soccer ball, for American readers) has a circumference of 70cm, a surface area of around 1,560cm^2, and an overall volume of around 5,792cm^3. A golf ball is much smaller, with a circumference of about 13cm, making its surface area 54cm^2 and its volume around 37.1cm^3. By volume, therefore, you could fit 156 golf balls into the same space as a football, but the total surface area of all those golf balls would actually be 8,424cm^2, which is a substantial two and a half square metres more than the football. Now imagine that on a nano scale, with 156 mini golf balls inhaled. Those 156 mini particles would weigh about the same as the one big particle in terms of gram per cubic metre – the preferred measurement for PM2.5 – but would nestle into way more lung tissue, causing more inflammation.

'The surface area difference is huge, and the surface is where all the toxic components lie,' David Newby reiterates. 'Levels of pollution at the roadside or in traffic are very much higher, and concentrated, than the background monitoring stations would suggest.' The majority of particles coming out of cars, especially the modern ones with good filtration systems, 'are less than PM0.1', says Newby.

Jim Mills, MD and founder of the pollution monitoring company Air Monitors, tells me that his industry is increasingly looking at isolating and identifying these smaller particles: 'We're moving at the moment to devices which now count and size the particles and by doing so it can tell us a lot about the "lung deposited surface area" – LDSA is a term that you're going to hear much more of. So that tells us how much capacity the surface has for having nasty things on it, it tells us how many particles there are which is important, because the particles are becoming smaller.' To explain just how small nanoparticles are compared to the PM10 we can see in black smoke, Jim says, 'imagine a PM10 particle weighs one gram – it doesn't, it's much, much less, but let's just call it one mass unit or one gram – if you ask most people what a PM1.0 particle weighs compared to that their answer will be 0.1 grams, perfectly reasonable isn't it? Except that it's not. If a PM10 particle weighs a gram, a PM1.0 particle weighs a milligram [a thousandth of a gram] and a PM0.1 particle, which is still bigger than most diesel particles coming out of an exhaust, weighs a millionth of a gram ... they've got virtually no mass – you would need a million PM0.1 particles to make the same mass as one PM10 particle ... as you go down in diameter by a factor of 10 you go down in mass by a factor of 1,000. So, what it boils down to is the particles coming out of modern diesel cars essentially have no mass, they're a million times less massive than a PM10 particle.' But yet PM10, and at best PM2.5, remains the core of most particulate matter legislation, regulation and monitoring? Yes, says Jim: 'The politicians say "Ah, but PM2.5 and PM10 include everything down to zero". Yes, but the mass isn't what we should be measuring, it's the number and the surface area of those particles.'

From his work on the London Atmospheric Emissions Inventory, Dr Beevers also finds that 'large numbers of very small particles don't really influence the mass very much ... The numbers of particles that you breathe in is really very localised, even more so than NOx, because vehicles emit really large numbers of very small particles ... The fall in number of particles from, say, a roadside to a background location is very, very steep.' The same is true of aircraft emissions. A 2016 study at LAX airport found lower PM2.5 concentrations than the researchers expected, despite the huge volume of emissions coming out of the planes. But when the researchers investigated by particulate number, they found excessively high levels – the planes' emissions were largely made up of nanoparticles that barely register in PM2.5 mass measurements. Further downwind they reacted within the atmosphere and formed into larger particles, contributing to high PM2.5 mass readings in the local area. They came out of the aeroplane as golf balls, and by the time they hit the surrounding area they had bunched together into footballs.

As well as having a greater surface area, nanoparticles are also able to penetrate deeper and further into the human body. In 2017, Newby's Edinburgh-based team set out to prove once and for all how this happened. They ground gold down into nanoparticles, got lab mice and human volunteers to breathe them in, and watched where in the body they ended up. It had long been an open question as to 'whether particles can directly cross into the bloodstream', he explains. 'We thought, what particle can we use that's not going to cause any damage to anyone but that we can detect in the body? We thought of various approaches including radioactive labelling, but ultimately we hit on the idea of using gold because you don't have any gold in your body, normally, and it is inert – the reason people make jewellery out of it is it doesn't oxidise, it doesn't cause any harm. We got a machine over from the Netherlands able to generate a cloud of gold particles, and generate different sizes. The small particles were 20 nanometres [PM0.02] up to around 50–60nm [PM0.05–6].' It was the mice's turn first. 'The smaller particles got down into the bloodstream much more readily, and the larger particles hardly got through at all. It seemed to be a cut-off point of around 30nm [PM0.03].'

The smallest particle my Egg PM2.5 counter can recognise is PM0.3, or 300nm, whereas Newby is talking about particles a tenth of that size (or a thousandth of the mass) and smaller being the ones that reach the bloodstream. 'In the end we tried a range from 2nm [PM0.002] to 200nm [PM0.2], and it consistently seemed to be around 30nm that it stopped crossing over from the lungs into the bloodstream … So, having done that on animals, we then went on and did it in people. We took healthy volunteers and exposed them to some gold particles in a chamber with a mask. We then got them to cycle a little bit to increase their breathing. And then we tracked their blood and their urine to see if we could detect the gold – and sure enough we could.' Gold in the volunteers' bloodstreams was detectable as soon as 15 minutes after exposure in some subjects. There was also evidence from the mice that the nanoparticles would cluster around areas of existing inflammation, or fatty deposits in the arteries, leading Newby and the team to conclude this could be the mechanism for inhaled nanoparticles to trigger acute cardiovascular events – namely heart attacks and strokes, caused by arterial blockages. If you imagine the artery as the road, and the prior inflammation or fat as the scene of a car accident, then the inhaled nanoparticles are the cars piling up behind it and causing the traffic jam. It had long been theorised that ultrafine particles smaller than PM0.1 could make their way almost anywhere in the body, including the lymph nodes, spleen, heart, liver and even the brain. Now Newby's Edinburgh team had proved it.

It's nanoparticles, then, and the total number of them that we breathe in, that pose the greatest danger to our health. To put it glibly, our bodies can deal with a bit of salt and sand in the air. But the vast majority of nanoparticles below 30nm, the ones that swim around our arteries and cluster around our cholesterol, don't come from transboundary pollutants. They come from our roads. More specifically, they come from the cars and vehicles burning fossil fuels – ancient lakes of decayed biological matter – for propulsion.

CHAPTER FOUR

No Smoke Without Fire

At this stage in my research I became completely side-tracked by a persistant question: have humans always lived with air pollution, or is it a modern phenomenon? The answer wasn't as straightforward as you might imagine. Consider this chapter a pause in the building drama. Let's take a breath, sit round the campfire and commune with our ancient ancestors a while. When was the year dot for air pollution? And did it do our ancestors any harm, or have we just become modern-day wimps?

'You can recognise a Neolithic site a mile off,' says Professor Paul Goldberg. His hands freeze in mid-air on the Skype call. Speaking from an archaeological dig in rural France, in a cave called Pech de l'Azé, his internet connection is, understandably, not great. But I can still hear him. 'A Neolithic site', he says, 'is just completely grey with ash.' Goldberg, a professor emeritus of geoarchaeology at Boston University, was involved in the discovery of the earliest known site where *Homo erectus* may have controlled fire. In Wonderwerk Cave, South Africa, Goldberg and his team found remnants of ash dating back to one million years ago. I'm expecting to talk to him about this marking 'year zero', the point after which fire and smoke became something quintessentially, enduringly human. But Goldberg and others in his field are not convinced. The archaeological evidence shows that rather than discovering fire and then using it every day, fire use actually comes and goes for vast periods of early human history. Some humans lived their whole lives without it, even after it was a known technology. It is not until we reach the Neolithic, as recently as 12,000 years ago, that sites of human activity are instantly recognisable by the piles of ash they/we left behind. So, what was happening in the roughly 988,000 years in between?

'Wonderwerk is a one-off!' exclaims Goldberg, who can't help but exclaim most points. 'The most convincing one is still Qesem Cave in Israel, which is only 300,000 to 400,000 years old. There you have repeated ashes one on top of another – it is clear, repeated fire … in Europe and the Middle East, there seem to be different fire records.' A paper from the University of Liverpool in 2016 agrees that 'despite the increasing numbers of [European Palaeolithic] fire sites, their *relative* scarcity is still notable.' Evidence of regular fires in any European site older than around 400,000 years is almost non-existent. If African hominins had mastered fire use before migrating into Europe, why didn't they bring that technology with them?

In 2015, Goldberg was one of 17 researchers invited to a symposium in Portugal called 'Fire and the Genus *Homo*', which focused specifically on the beginning of fire use and its role in our evolution. Among the 17 was Dennis M. Sandgathe, lecturer in archaeology and human evolutionary studies at Simon Fraser University in Canada and the University of Pennsylvania. Having spent much of his earlier career specialising in Stone Age tools, in recent years he too became side-tracked by the questions surrounding early human's use of fire. 'For the period prior to about 400,000 years ago we have maybe half a dozen sites with any evidence for fire,' Sandgathe tells me, speaking from his home office in Canada. 'And it's not overwhelming at any of these sites – a few tiny fragments of burned bone, a reddened patch of sediment … an incredibly small number of potential examples of fire use for such a long period of time … I am not convinced that any of these sites are related to human behaviour – they might be, but it is just as likely, perhaps more likely, that they are mostly natural fire residues – certainly none of them are demonstrably the product of people using fire. For the period after around 400,000 years ago we do have clear evidence of people using fire,' and again he points to Qesem Cave in Israel.

The difference between Qesem Cave and all earlier sites is the presence of a hearth – a purpose-built (in this case

4m-square) fire pit. This site, in a limestone ridge 12km (7.5 miles) east of Tel Aviv, Israel, is the current earliest-known site where humans first sat, regularly, around a fire. While the cave was occupied from 420,000 to 220,000 years ago, the hearth is dated from 300,000 years ago. Amazingly, given its age, this site also offers our first window into the effects of regular smoke inhalation. A 2015 paper by Karen Hardy, Universitat Autònoma de Barcelona, studied eight hominin teeth from Qesem Cave for evidence of potential smoke inhalation. She found 'the earliest evidence of exposure to potential respiratory irritants', including micro-particles, in human history. The presence of small micro-charcoal fragments in the teeth, particles up to 70nm, entered the mouth through breathing, not eating, 'indicative of fire and these suggest a smoky atmosphere inside the cave'.

Is this 300,000-year-old hearth in Israel, then, our year zero, after which humans and fire would always co-exist? It's still not that simple, says Sandgathe. He and Goldberg have both recently turned their attention to the use of fire by Neanderthals, the Eurasian contemporaries of the Qesem Cave dwellers until their extinction (or assimilation) around 40,000 years ago. 'Everyone, including us, assumed Neanderthals always used fire and likely knew how to make it,' says Sandgathe. 'Then we excavated ... and found that, while they were definitely using fire during some periods, there were long periods of time where they weren't using fire at these sites ... and the layers with little fire evidence are, strangely, associated with cold periods. These layers still have tonnes of artefacts and bones so they are still inhabiting the sites during these periods – it's not like they stopped using the sites.' There are at least seven sites in south-west France that match this pattern of reduced fire evidence during cold periods. In layers that date to the last interglacial period between 130,000 and 75,000 years ago, fire was being used quite frequently by Neanderthals, using fully intact hearths. In the layers at the same sites during cold periods, from 75,000 to 40,000 years ago, fire use seems to have disappeared. These archaeological layers are still stuffed with stone tools and butchered animal bones, but none of them are burnt, and there is no hearth. It would suggest that Neanderthals had forgotten how to make fire, were it not for the fact that they had already mastered advanced fire techniques. As well as constructed hearths, lumps of preserved pitch were found in a Neanderthal site in Germany in 2001, dating from roughly 80,000 years ago. Pitch, probably used as a wood glue, can only be made from tree bark kept at a constant high temperature in a controlled fire for several hours.

'So, there's lots of interesting questions … maybe they didn't need fire?' says Goldberg. 'And this is much younger [than Wonderwerk] – it's at the end of human history, not at the very beginning.' So, fire as part of daily human life isn't necessarily a given? 'No! That's the thing. Neanderthals here in France used fire 80,000–90,000 years ago, and then it got cold, and they stopped. We were talking about this idea over breakfast this morning at the dig site – the idea that "oh, we discovered fire, and it took off, and everybody was using it" – but it's not true. It's simply not true.'

What does this do to the 'cooking hypothesis', that cooking food led to our unique brain growth relative to other animals? 'Well it doesn't support it, put it that way,' says Goldberg. 'If that's really true, then we should find fire everywhere. And we don't.' The alternative theory, therefore, is that for millennia – in fact, if it is right, for the majority of human existence – we've been happy to tenderise, pickle or ferment food, rather than cook it over a fire. There is even a possibility that many early humans didn't want to live in close proximity to fire. I think this is a fascinating challenge to our notion of what it is to be human. I am not suggesting we return to pickling and slow-chewing our food; but what I am suggesting, perhaps provocatively, is that maybe we don't need fire as much as we think we do? Maybe we can, like our ancient ancestors, pick and choose when we use it?

From the agricultural revolution during the Neolithic period onwards, fire did become central to human civilisation. People started living together in townships, and they managed crops and livestock from fixed abodes. The hearth became not just central to every community, but central to every home. 'You can spot a Neolithic site a mile away', repeats Goldberg, when he reanimates over Skype, 'because all the sediments are grey and ashy ... people started sitting down and became sedentary, the whole structure of society changed ... probably about 12,000 years ago – it's just major change ... From the point of view of pollution, I would say that's when it starts – in the early Neolithic.'

Dr Hans Huisman, archaeologist at the Dutch Ministry of Culture and previously a lecturer at Leiden University, is a micromorphologist specialising in the Neolithic. He was part of a study team at Swifterbant, a unique Neolithic site preserved under the Dutch wetlands. For centuries the Netherlands has used dikes to drain the coastal flatlands and turn them into 'polders' fit for agriculture. When a wetland at Swifterbant, 50 miles east of Amsterdam, was drained in this way in the 1960s, a complete Neolithic landscape was revealed dating from around 6,000 years ago. At that time, when global seawater was lower, it was a settlement on dry land amid tidal creeks. Among the usual pottery, bones and tools, there was also plenty of burnt stuff. 'Classical Neolithic settlements of the first agriculturalists', says Huisman, 'are rich in ceramics, there is burnt material, there are flints everywhere, it's very much concentrated ... In this region the sites have thick, black layers full of archaeological remains that consist mostly of burnt plant material, because this was preserved below water and silt, rather than farmed fields or building sites. 'What's interesting [at Swifterbant] is you not only have all the finds, but they are also in exactly the same position where they were 6,000 years ago – so the preservation is really exceptional.' There were also plenty of well-preserved hearths. 'Most of these buildings were combined with people and cattle often in the same building, and you would have two-thirds of the space for cattle and one-third for the people. You often find one hearth in the people's section and a second one in the centre of the building. In Neolithic conditions, you have daily contact with smoke ... in fact fire is burning at least a little bit all day ... from the Swifterbant site several striker lights have been found – flints that are specially made to make fire … I think everybody must have been able to make fire and use fire, and you see them using it on the settlement and in the landscape, for food, for heat, for waste disposal, for producing artefacts.'

Pre-dating the mining of fossil fuel, the common fuel of the time would have been wood, 'and if no wood was available, in this period, it was probably animal dung,' says Huisman. 'There is some discussion in Palaeolithic settings whether they also used bone as a fuel – fresh bone is full of fat and that would burn.' Dried dung, a common fuel at the time and still in many parts of the world today, is very smoky. 'Even if you have it well dried, it is still very smoky,' says Huisman. 'You don't get a real flame – it sort of smoulders … I know a guy who has been doing experiments on different types of fuel in a reconstructed medieval farm building, and the wood fire and peat fire did not cause him many problems, but when he used the animal dung he had to get out because the whole building was full of smoke!' He laughs, adding, 'A wood fire you can keep burning in the night as you sleep; a dung fire, you have to put out.'

Another Dutch archaeologist, Dick Stapert, theorised in the 1990s that the evolution of a complex language, involving abstract concepts, may be thanks to daily gatherings and story-telling around the hearth, and even stimulated the development of the arts. Fire also started to appear in proto-religious contexts. People began to be buried with their 'lighters' – flint and pyrite kits – presumably for use in the afterlife.

Then, around 7,000 years ago, our mastery of fire sparked arguably the first true industrial revolution: metalworking in copper, bronze and then iron. Decorative copper beads have been found in Çatalhöyük, a large Neolithic settlement in Turkey, dating from around 7500 BC to 5700 BC. Later, bronze was made by heating the ores of copper and arsenic together, producing the first toxic industrial emissions alongside prized bronze tools and ornaments. The earliest known example of metallic lead is a metal figure recovered from the Temple of Abydos in Upper Egypt, from around 6,000 years ago: lead requires not just extracting ore from rock, but smelting – a high-temperature process that burns off the sulphur, binds the ore with oxygen, and reacts with carbon, typically charcoal.

Smelting lead ore to extract silver came with the ancient Greeks, around 1350 BC, before the Roman Empire adopted and massively increased the technology. The world record levels of airborne lead emissions produced during this period would not be topped until the Industrial Revolution a couple of millennia later. Around the time of the birth of Christ, silver mines were producing 80,000 tonnes of lead slag a year – at least 1 per cent of which were particles small enough to mix into the air. Modern ice-core sampling has revealed that around 400 tonnes of lead particles fell on the Greenland ice cap during the 800 years of the Roman Empire (though this is just 15 per cent of the lead that fell during the 60 years of leaded petrol in the twentieth century). The smell of the air also changed. The Roman chronicler Lucius Annaeus Seneca the Younger wrote in AD 61, 'As soon as I had got out of the heavy air of Rome, from the stink of the chimneys and the pestilence, vapours and soot of the air, I felt an alteration to my disposition.' One of the few embalmed mummies discovered from ancient Rome, known as 'the Grottarossa mummy', shows severe anthracosis – accumulation of carbon in the lungs from repeated exposure to smoke – in a girl who only lived to the age of eight.

I visited the British Museum in late 2017 hoping to see some of the evidence of ancient air pollution first hand. I had arranged to meet Daniel Antoine, Curator of Physical Anthropology, responsible for all the museum's human remains, and his PhD student Anna Davies-Barrett, under the public glass atrium. Daniel leads us to a side door, turns a key, and takes me behind the scenes. Suddenly the architecture is less Norman Foster, more 1970s council office. We go up a narrow set of stairs, down a corridor, to a door that opens into a surprisingly small room. Two almost complete skeletons are laid out on tables, their bones an orange-brown from desert soil. They are both medieval Sudanese women, who lived next to the Nile river. Volunteer researchers are cleaning their bones and teeth with dry brushes.

These skeletons are around 1,000 years old and are just two of nearly 1,000 skeletons donated by the Sudanese National Corporation for Antiquities and Museums to the British Museum in 2007, excavated prior to the building of the Merowe hydroelectric dam on the Nile. Well preserved by the arid desert environment, the skeletons reveal how these societies lived from 1750 BC to AD 1500. Anna is specifically studying the prevalence of chronic respiratory disease which, in the absence of any remaining lung tissue, means she must look at the bones themselves. 'If you have inflammation from a respiratory disease you might expect to have new bone forming in the sinuses, or on the inside layers of the ribs,' she tells me. Lower respiratory tract diseases can cause inflammation near certain points in the ribs, causing new bone growth. Similar rib lesions were found in the skeletons of ancient Romans buried by Vesuvius in AD 79, believed to be caused by long-term smoke inhalation.

'There's some partial natural mummification,' Anna points out. 'You can see some skin and ligaments here,' she indicates some papery fragments on the skull. I ask what they know about this individual. 'You can see from the pelvis it is a female,' says Daniel. 'And from the wear and tear of the joints ... she was probably a middle adult when she died, so about 35 to 50 years of age.' They can also see evidence of disease, including fractured bones, a twisted spine probably caused by tuberculosis, and, somewhat less scientifically, 'terrible teeth'. In coming days Anna will also check for sinusitis, using an endoscope – a tiny camera on the end of a thin tube, more commonly used in surgery – to avoid any damage to the skull: 'I often refer to sinusitis as the barometer for air quality,' she tells me. 'It is a fairly good indication, if you have really high rates in a population, because it is your first barrier of defence between your body and the air you breathe ... if there was inflammation, it is really suggestive that something in the air that you are breathing in is causing that.'

Sinusitis is the inflammation of the lining of one or more of the nasal sinuses – the four pairs of air-filled cavities in the skull above and around the eyes and nose. If the inflammation is chronic and ongoing, then the pressure causes new bone growth – vital tell-tale signs for the future archaeologist. Anna tells me that 'previous bioarchaeological studies looking at the ribs and the sinuses have found ... very high rates of sinusitis that may have been caused by metalworking, suggesting that fine particulates in the air from metalworking was causing the men to have very high rates of sinusitis.' While she hasn't yet published her findings when we meet, she has found a high rate of sinusitis among the Sudanese skeletons she has examined thus far.

'Human remains and their study is allowing us to see evidence of our past human health that other sources don't provide, whether it's written texts or material culture,' says Daniel. 'It is only by studying human remains that we can get an insight into past states of health ... and they are very relevant to today. Cancer and cardiovascular diseases are not new diseases. And by looking at where and how they are being expressed, maybe we can get a better sense of what circumstances and conditions lead to an increase in disease prevalence ... whether just living in a certain environment can lead to a higher rate of respiratory disease.'

Daniel, who also looks after the Egyptian mummies in the British Museum, tells of similar findings within that collection too: 'For Egyptian mummies, they would remove the lungs as part of the embalming process, sometimes they would put them in canopic jars [used during mummification to preserve important body organs]... Some research has looked at the lungs in the museum collection canopic jars and found evidence of silicosis [silica or sand inflammation, most likely from dust storms] and anthracosis, exposure to carbon, and finding particles of both carbon and sand in the lungs ... What's interesting is that the mummified people would have been the very wealthy people. And even in them, who are not obviously involved in specific work activity, we are finding evidence of respiratory diseases and conditions.' The next planned study, he says, is to use CT scans of the mummies to look for evidence of atherosclerosis – the fat in arteries that leads to strokes, and around which, as Newby discovered, inhaled nanoparticles can accumulate.

When ancient Greece and Rome held sway, wood remained the primary fuel source. But then came coal. While there are recorded uses of surface-level coal in ancient China and Rome, there is little suggestion of coal being a primary fuel until the heyday of the eleventh-century Chinese Sung capital Kaifeng (500km or 310 miles south of Beijing). Kaifeng is thought to have been the first city in the world to convert its energy supply from wood to coal thanks to river and canal transport that gave it direct access to emergent coal mines. At its peak Kaifeng was a true mega-city with almost one million inhabitants – probably the largest city in the world at the time – with all its cooking and heating fuelled by the new black gold. This early urban smog proved short-lived, however. The city was sacked by Jin Dynasty troops in 1127 and ravaged by a double whammy of Mongol troops and the plague in the following century. It dwindled to little more than a village. The industrial potential of coal, however, with an energy density (the amount of energy stored in its mass) twice that of wood, was now an open secret. By the thirteenth century, coal mining was well established in Europe, and coal was carried by boat into the heart of London.

The first documented record of British air pollution appears when Queen Eleanor, wife of Henry III, cut short her visit to Nottingham in 1257 because of the city's overwhelming coal fumes. Edward I, her son, attempted the first environmental regulation to control it, setting up a commission in 1285 to come up with a solution. His subsequent attempt to ban coal, however, didn't have much effect and was soon rendered obsolete when – as with Kaifeng – the Black Death ravaged the country, killing off a quarter of the population; forest began to reclaim the abandoned farmland as entire farming communities were wiped out by plague, and wood became cheaper and more abundant again than coal. Professor Peter Brimblecombe, whose 1987 book *The Big Smoke* records the history of air pollution in London, comments that this sequence of events has been repeated many times since: rapid population growth, urbanisation, increases in population density, fuel shortages, followed by embracing new fuels which turn out to be much more polluting than their predecessors.

Two centuries later, the pattern was repeated: the population recovered, wood and charcoal became scarce, and coal – specifically sea coal, a low-quality, smoky coal from the sea floor found in abundance around the coast of Scotland – returned. Once again, the monarch – this time, Elizabeth I – complained of being 'greatly grieved and annoyed with the taste' of coal smoke. Shakespeare's 'man of the people' character Falstaff bemoans the 'reek' of coal-fuelled lime-kilns. Brimblecombe estimates that coal imports into London between 1580 and 1680 increased 20-fold. Tudor houses that still stand today typically have beautifully ornate, but comically tall, chimneys; like trees competing for canopy space they were trying to rise taller than each other, wafting their own foul smoke as far away as possible.

Over 400 medieval skeletons from two burial sites in the UK were examined for sinusitis in a 1995 study, in much the same way as I had seen at the British Museum. In Wharram Percy, an abandoned medieval agricultural village in Yorkshire, the residents' lives would have been far from smoke-free, with charcoal, coal and dung probably burned to heat their homes. But St Helen-on-the-Walls, an urban parish in the city of York, housed poor workers who lived with the same pollution at home plus the industrial emissions from the foundry, apothecary, tanning and brewery factories they worked in. Of the skeletons from the village with sinuses preserved, 39 per cent had evidence of sinusitis; from the urban site, over half (55 per cent) of the individuals had sinusitis, a 12 per cent difference which researchers from the University of Bradford attributed to 'industrial air pollution'.[1]

In the Americas, toxic air pollution arrived with the Spanish conquistadors. Ice-core samples taken from the Quelccaya ice cap in Peru have traced ancient metallurgy and mining in South America back to AD 798 and the sprawling Inca Empire, but the levels of pollution were low, from metal smelting undertaken as a relatively minor cottage industry, using small furnaces. After the Europeans came in the sixteenth century, however, they brought with them the technique of extracting silver using liquid mercury. The toxic dust it created fell all over South America. Some describe sixteenth-century Bolivia, where the Potosí silver mine was the largest in the world at the time, as the start of the Anthropocene – the geological era in which human activity began to have a significant impact on the natural world. The air would never be the same again.

Less than a hundred years later, the diarist John Evelyn chose the smoke-filled air of London as the subject for his new-found gentleman's pursuit: science. A founder member of the Royal Society of Science, in 1661 Evelyn undertook what is thought to be the first scientific study of air pollution. Addressed to King Charles II, the pamphlet, entitled *Fumifugium, or The inconveniencie of the aer and smoak of London dissipated together with some remedies humbly proposed by J.E. esq. to His Sacred Majestie, and to the Parliament now assembled*, appeals to the King's ego as well as his intellect: 'Your Majesties only Sister, the now Dutchesse of Orleans … did in my hearing, complain of the Effects of this Smoake both in her Breast and Lungs, whilst She was in Your Majesties Palace,' wrote Evelyn. 'I cannot but greatly apprehend, that Your Majesty (who has been so long accustom'd to the excellent Aer of other Countries) may be as much offended at it, in that regard also, especially since the Evil is so Epidemicall; indangering as well the Health of Your Subjects, as it sullies the Glory of this Your Imperial Seat.' Evelyn foresaw health impacts that would only be proven centuries later, in stating: 'Aer that is corrupt insinuates itself into the vital parts immediately … passing so speedily to the Lungs, and virtually to the Heart itself.' He also had the backing of fellow gentleman scientist Sir Kenelm Digby, who took note of Londoners dying from 'pulmonary distempers, spitting blood from their ulcerated lungs'.

Among the remedies Evelyn proposed was the public planting of shrubs and flowers on a grand scale: 'that all low-grounds circumjacent to the City, especially East and South-west … be elegantly planted, diligently kept and supply'd, with such Shrubs, as yield the most fragrant and odoriferous Flowers'. Shrouded in jasmine and lavender, London's air could be the envy of the world. Charles II actually agreed to it; sadly, Parliament did not, and the Great Fire of London just five years later saw Evelyn's planting scheme plummet down the list of priorities. He submitted a grand plan to rebuild the devastated city in line with his principles, with smoke-emitting industries banished downriver, but it was rejected – as was Christopher Wren's – in favour of simply rebuilding the city where it had stood.

By the Industrial Revolution, however, even Evelyn's wildest shrubbery scheme wouldn't have stood a chance. The English astrologer John Gadbury noted an increase in London's fogs in his weather diary from 1668 to 1689. By comparing Gadbury's diary with the official records of deaths, Peter Brimblecombe found that when a 'Great Stinking Fog' (Gadbury's expression) appeared, the number of deaths in the city doubled. At that time, London had a population of roughly 500,000, many of whom still lived and worked within the original 'Square Mile', encircled by Roman walls. Records of atmospheric CO_2, methane and nitrous oxide in the modern ice-core record start to rocket upwards from the point that, as the authors Seinfeld and Pandis suggest, coincides 'more or less with the invention of the steam engine in 1784'[2]. A tombstone in Kensal Green Cemetery commemorates a certain 'LR, Who died of suffocation in the great fog of London 1814'. This particular fog, which lasted from 27 December 1813 to 3 January 1814, was reported by *The Scots Magazine* as being full of the 'smoke of the city; so much so that it produce[s] a very sensible effect on the eyes, and the coal tar vapour [was] … distinctly perceived by the smell'.

By 1860 London had grown to over three million inhabitants and had sprawled to new suburbs, many far from the River Thames, such as Clapham, New Cross, Tottenham and Walthamstow, all connected by steam-powered trains burning coal. Industries belching coal smoke were now in the heart of every industrial town and city, and every home within them was heated by coal. London became famous for its black umbrellas – the only colour that didn't show the blackened rain. It was the biggest city in the world from 1831 to 1925, and its skyline, filled with giant chimney stacks and black clouds that never went away, had set the blueprint for the modern world. The northern industrial city of Sheffield, believed to have produced 90 per cent of the world's steel at the time, was described simply as 'black' in the travelogue *Rural Rides* by William Cobbett in 1830: 'All the way along, from Leeds to Sheffield, it is coal and iron, and iron and coal. It was dark before we reached Sheffield; so that we saw the iron furnaces, in all the horrible splendour of their everlasting blaze.'

It wasn't until 1905 that someone came up with a catchier name than 'Great Stinking Fogs'. On the opening day of the Conference for Smoke Abatement held in London in December 1905, Dr Des Voeux suggested combining the words smoke and fog to become 'smog'. It was a throwaway line, one that the assembled delegates, probably including Des Voeux himself, thought little of compared to the important proceedings on the three-day itinerary (such as the 'Final Report of the Royal Commission on Coal Supplies'). But the press picked up on it, and the word 'smog' quickly entered the English language on both sides of the pond.

In the US, Dr Devra Davis is one of the most important names in the modern fight against air pollution. A leading academic with numerous papers to her name on the link between health and pollution (long before it was a topic that attracted the funding bodies), she has lectured at universities including Carnegie Mellon and Harvard, medical schools in Jerusalem, Turkey and London, and advised on government boards nationally and internationally. She also happened to be a child in Donora in the 1940s, during one of the world's most infamous air pollution episodes.

In the first half of the twentieth century, Donora, a small American mill town in Pennsylvania, was well accustomed to dirty air. The sun often didn't shine for days, blocked out by fumes from the steel mills, coke ovens, coal stoves and zinc furnaces, trapped in the valley by the surrounding hills. Davis tells me that the chemicals in the air could produce 'astonishingly beautiful sunsets'. But Friday 28 October 1948 was anything but beautiful. A pocket of air pressure caused a temperature inversion, with a layer of warm air trapping the colder boundary layer close to the surface. The pollutants were stuck in a stagnant layer of air only a few metres from the ground. One eye witness, the town's attorney Arnold Hirsh, told Davis that he saw a steam engine on the tracks as the smog formed: 'It issued a big blast of black smoke that went up about six feet in the air and stopped cold. It just hung there ... in air that did not move.' The air soon became a thick, yellow soup of sulphuric acid, nitrogen dioxide and fluoride gas, mostly emitted from the zinc smelting plant. It lasted for four days. The company operating the Donora Zinc Works finally ordered the plant to shut down at 6 a.m. on Sunday morning; the next day, the smog had gone. Twenty people died during the incident, and 7,000 (almost half the town's population) were hospitalised, with a further 50 dying soon after. Autopsies revealed that the inhalation of fluoride gas – released from the smelting of fluorspar in the zinc factories – was the primary cause of death.

In the years that followed, the townsfolk never talked about the disaster, recalls Davis. The plants stayed open and life continued as if it was any other small town in America, with neat lawns, white picket fences and pink curtains – the difference being that the grass rarely survived, the picket fences would quickly turn black, and Davis's mum preferred venetian blinds to curtains because 'they were easier to wipe'. Hairdressers would go from home to home 'to take care of the little old ladies', remembers Davis, 'but these were women in their fifties – they were bedridden because it was hard for them to go up and down the stairs because there was so much heart disease ... The fact that the skies were brown and we didn't see the sun for days at a time – that was just normal. Especially in the fall. And the dirt that was always on us, from being outdoors – we used to call it the 'Donora measles' because you'd get black spots.' While she was only two at the time of the disaster, she later interviewed a number of survivors, including her family, for her book *When Smoke Ran Like Water*, published in 2002. 'I interviewed a man who had just come back from the war in Europe, and was in excellent physical condition – he talked about gasping for breath during this episode … one of the first clues that something was wrong was that the funeral homes ran out of caskets, and the florists ran out of flowers, and drugstores ran out of drugs. People knew that something was going on.' A high-school football game went ahead despite the kickers not being able to see where the ball ended up. Mid-game, a Donora player was told over the tannoy that he needed to go home immediately. By the time he got home, his father, an iron worker, had died. The mill owners American Steel and Wire Works never admitted any responsibility, calling the chemical-filled fog 'an act of God'. The plants stayed open until 1962.

For as long as humans have burned things, then, air pollution has been a killer. But as our industrial practices became more advanced, the wood smoke of our ancient ancestors in Wonderwerk and Qesem morphed into something entirely different. The daily proximity to smoke persists, but rather than from an open hearth it pours from thousands of combustion engines and industrial furnaces. Modern economies began discovering and burning more and more chemical compounds, and the smoke became deadlier. A physician from the University of Cincinnati, Clarence A. Mills, wrote in the journal *Science* in 1950, 'Let us hope that the Donora disaster will awaken people everywhere to the dangers they face from pollution of the air they must breathe to live.' His hopes were not realised. Instead, by the early twenty-first century, humans have unwittingly turned the air into the largest known environmental health risk. The pattern that Peter Brimblecombe identified reappeared once again: rapid population growth, urbanisation, increases in population and a new fuel that turns out to be more polluting than its predecessors. This time, the fuel was diesel.

CHAPTER FIVE

The Dash for Diesel

At the Millbrook emissions testing ground, 40 miles north of London, security guards take my phone and laptop and cover the cameras in thick red tape. As a journalist I've visited many security-sensitive sites before, from the Houses of Parliament to nuclear power stations, but this is a first for me. When I clear security and step through the gates, I find out why. Paranoia here is high, and so is the amount of money on display. The Millbrook Proving Ground, to give it its full name, is a rite of passage for any new car to make it from a mere prototype to a real-world, production-line model. It is the size of a small town, with emissions testing labs, crash-test centres, noise chambers, atmospheric chambers and 70km (43 miles) of test tracks. Logo-less cars drive around painted in zebra-like stripes to camouflage them from prying lenses, were I to surreptitiously remove a camera sticker.

Millbrook is an independent business – all its activities are funded by the car companies (or, as they call them in the industry, 'OEMs' – Original Equipment Manufacturers) who test their new cars and engines here and pay handsomely for the privilege. By the side of one test track I spot a gleaming Aston Martin showroom nestled between the trees. Its high-net-worth customers can test drive their bespoke vehicles on the same track where James Bond sequences have been filmed.

I was invited to visit Millbrook alongside members of the London Assembly – the elected body that scrutinises and advises the Mayor's Office – here to learn more about the emissions testing process and check up on the fleet of London's famous red buses that are tested and accredited here. And I, as the only member of the press tagging along, am given a surprisingly long leash.

When Phil Stones, Millbrook's head of emissions and fuel economy, sits us down for a presentation, two things raise eyebrows. First, nitrogen dioxide (NO_2), the EU legal limit of which is regularly breached in London, is not part of the Euro vehicles emissions test. NOx (all oxides of nitrogen combined) is, but the proportion of nitrogen dioxide within that can vary – and the proportion, Phil tells me later, is going up. Second, PMs from tyre and brake wear, despite being known to form a significant percentage of overall PM from traffic, are also not covered by the Euro standards, and therefore not of interest to the manufacturers. 'There is no legal standard – no legal requirement,' says Phil. He adds, not for the last time that day, that manufacturers will only build to meet the regulatory standards, and no more.

When it comes to PM2.5 and nitrogen dioxide within towns and cities, by far the greatest single source is the modern automobile. Airparif's annual reports on the pollution in Paris include 'hotspot' maps for nitrogen dioxide and PM2.5, showing red for high pollution, down through yellow and then green for low pollution. On the maps, the city's roads glow bright red against a yellow-green background. These markings clearly show that urban emissions originate from the roads. In maps that zoom out to show the wider Île-de-France region, Paris is a blotch of red in a sea of dark green, with the occasional yellow vein of a motorway running across it. Traffic accounts for 65 per cent of the NOx and over half the PM2.5 inside the city of Paris. And the vast majority of that comes from diesel. In Paris, diesel vehicles account for about 50 per cent of the traffic, but 94 per cent of the NOx and 96 per cent of the transport-derived PM10.

The exhaust from internal combustion engines produces two kinds of particles: secondary and primary. But the ratio and total number differ, depending on whether it is a diesel or petrol engine. Professor Simone Hochgreb, formerly of Princeton and MIT and now at the University of Cambridge works almost exclusively on engine emissions: 'Suppose you start with 100 per cent of fuel, and say only 98 per cent of it burns,' she explains. 'This means you end up with 2 per cent coming out either as the original fuel or as a messy mix of partly burned stuff, VOCs or particulate matter ... This is what's called "incomplete combustion". In gasoline engines, incomplete combustion arises primarily from fuel near walls and crevices near the piston when the engine is warm, and from spraying a bit too much fuel when it is cold so that the engine starts. In diesel engines, it arises because you inject the fuel separately from air, and depending on the operating conditions, there is not enough mixing or not enough time to completely convert it into CO_2.' The diesel engine and the gasoline engine therefore work on two very different principles. 'There is a trade-off in NOx and PM for most types of engines, certainly for diesels,' says Hochgreb. 'If one goes down, the other goes up ... the simple idea is that a diesel engine is very efficient because it can operate at very high pressures, which also produce high temperatures, which also produce NOx. But in order to use those high pressures and temperatures, you can't inject the fuel too early because otherwise it will go boom. So, in diesel engines you inject when the air is hot – but now the fuel is not completely mixed with the air, so it will not burn completely, and it will produce the PM, the soot, which comes from incomplete combustion ... the PM level is bad news.'

* * *

I was at a rural wedding in Dorset, southern England, when the Volkswagen diesel emissions scandal broke in September 2015. During the evening reception, in a marquee set in an apple orchard, I slipped outside for a breath of fresh air. I got into conversation with an uncle of the groom, which all too quickly strayed into work. 'You write for the *Financial Times*?' he said. 'What did you make of this morning's front page – Volkswagen, eh? It's got to be the biggest scandal since Enron.' I hadn't read that or any paper that morning – I had been trying, with varying degrees of success, to entertain my now 18-month-old daughter on the five-hour train journey to the wedding. So, he filled me in. VW had been caught falsifying the emissions data from its diesel cars, cheating test results to sneak under the radar of increasingly stringent regulations and appeal to eco-conscious consumers. As he spoke I gazed at the bonfire that flickered in the orchard, and thought of my mum's diesel VW Polo, and the low-carbon credentials she'd mentioned when she bought it.

But the story of how diesel came to dominate our roads began long before a VW engineer said: 'Hey, why don't we just cheat?!' A couple of decades before that malevolent lightbulb moment, ill-considered government policy had already released untold tonnes of diesel fumes into the air.

The Kyoto Protocol in 1992 required governments to reduce CO_2 emissions by 8 per cent from 1997 to 2013. Given the known global warming effects of CO_2, this was a very necessary step. The means of reduction, however, wasn't stipulated, and in Europe it led to the widespread adoption of diesel vehicles. Due to its relative fuel efficiency diesel can produce 15 per cent less CO_2 than petrol engines. The car industry, sniffing an opportunity to build vast amounts of new cars, lobbied the European Commission to promote diesel. It was pushing on an open door. In 1998 the EC issued a commitment to cut CO_2 emissions by 25 per cent in all new cars sold within ten years; the only available means of doing so was a switch to diesel engines. Most EU countries began to introduce tax breaks to incentivise consumers to buy diesel cars over petrol. In 2001–2 the UK began taxing vehicles according to CO_2 emissions. Cars with lower CO_2 emissions fell into cheaper vehicle excise duty (car tax) bands, which gave diesels a cost advantage, and fired the starting gun for what became known as the 'dash for diesel'.

It was a case study in effective government policy. With similar tax and fuel incentives rolled out Europe-wide to meet the EC ruling, the market share for diesel cars across the continent rose from under 10 per cent in 1995 to over 50 per cent by 2012, while the share of diesel in total fuel consumed reached 63 per cent. From 2001 to 2010, the proportion of diesel among all new cars registered in Norway rose from 13.3 per cent to 73.9 per cent and in Ireland from 12 per cent to 62.3 per cent. Buoyed by such success, European carmakers took their diesel models global. In India, diesel went from just 4 per cent of new car sales in 2000 to half of new car sales by 2016. When I visited the Central Road Research Institute (CRRI) in Delhi, Dr Niraj Sharma told me that diesel was initially highly subsidised for agricultural use:* 'It used to be that 90 to 92 per cent of all vehicles used to be petrol driven, and around 8 to 9 per cent were diesel,' he says. 'Now things have changed ... diesel vehicles started dominating ... the subsidy that was meant to be for the farmers, is mostly being used by the automobile manufacturers.' Diesel on the streets of Delhi, he says, emits 'more hazardous air pollution than the petrol-driven vehicles'.

In 2000, when these incentives started, even the most efficient diesel cars emitted over three times more NO_2 per kilometre, and ten times more PM. Most cars on the road were much older, emitting at least four times more NOx and between 22 and 100 times more PM than petrol engines. A conscious trade-off was therefore made by policy-makers to accept ill-health as a result of increasing pollution in our towns and cities, in order to meet CO_2 targets. Simon Birkett, a former London City banker who gave up his job to set up the campaign group Clean Air London in 2007, told a *Guardian* exposé in 2015 that 'even though the European commission, national governments and the car industry knew how dangerous diesel is, together they incentivised it and deliberately engineered a massive switch away from petrol – without any public debate.' The article further quoted a retired 'very senior' civil servant who recalled 'the health issue' as being a significant factor in departmental debate at the time: 'We did not sleepwalk into this ... everyone had to swallow hard.'[1]

Any claims that 'we didn't know how bad diesel was' ring extremely hollow. The International Agency for Research on

---

* Diesel is good for uses where you want high torque and low speed, such as tractors.

Cancer (IARC) had named diesel exhaust as a probable carcinogen back in the 1980s. In 1986, Dr Robin Russell-Jones, a lung expert who successfully campaigned against leaded petrol, gave evidence to a UK House of Lords select committee that diesel pollution was linked with asthma, cardiovascular disease and lung cancer. In 1993 a major report for the Department of the Environment by the Quality of Urban Air Review Group (QUARG) stated that diesel emissions were 'a potential health hazard', containing 'compounds known to be carcinogenic that may cause impairment of respiratory functions ... an increase in mortality and morbidity may be associated.' In 1996, a POST (Parliamentary Office of Science and Technology) report given to UK Members of Parliament to brief them on scientific developments warned of 'emerging evidence that fine particles in the air could be a significant contributor to respiratory disease and death ... fine particles from diesels and other sources may contribute to significant mortality around the world.' It continues, 'road transport is the biggest single source of particulates and diesel emissions are the dominant source ... There is also evidence linking exposure to diesel exhaust with higher rates of lung cancer.' The POST report again drew attention to the earlier findings of the QUARG report, just in case anyone missed it the first time, that any increase in the market share of diesel – which was around 20 per cent at the time – would 'inevitably make matters worse as the current technology diesel car emits far more particulate matter than the modern petrol car' with 'no current prospect of the diesel improving beyond the petrol car'.[2] When I read that, I genuinely got goose bumps because the information was there, in black and white. Ignorance was not a defence. But no one heeded the warning.

Peter Brimblecombe, who sat on the QUARG committee at the time, clearly remembers that 'the message of the group was that particles are the big problem and they're diesel particles.' Records received by the BBC in November 2017, after a two-year freedom of information battle, confirmed that ministers and civil servants in the government were well aware that diesel pollution was bad for air quality. Advice from the Treasury's tax policy section presented to ministers – the most senior members of the government – unambiguously stated: 'Relative to petrol, diesel has lower emissions of CO_2 but higher emissions of the particulates and pollutants which damage local air quality.'

Governments across Europe ignored all this. Promotions for diesel as the 'green' option continued unabated. There was some (very slight) method to this madness, however: the Euro emissions standards, which all new cars had to pass before sale in the EU, promised that cleaner diesel was just around the corner. The Euro 1 emissions standard began in 1992, requiring all new cars sold within the EC from that point to meet certain emissions criteria for NOx, PM, CO and hydrocarbons. The bar started off high, but the plan was always to bring it down a notch or two with each new emissions standard: Euro 2 came in 1996, followed by Euro 3 in 2000 and Euro 4 in 2005 (we are, at the time of writing, currently at Euro 6, introduced in September 2014). Due to the inherent differences in engine technology, diesel was allowed to emit higher amounts of NOx and PM than petrol, but the gap was due to get ever smaller. The car companies could repeatedly say, 'yes the emissions are high now, but just wait till you see our next model'.

Frank Kelly, who has been part of the government advisory Committee on the Medical Effects of Air Pollutants since the early 2000s, admits: 'There were reports saying, "if you make this decision based on benefit from CO_2 decreases there is the likelihood that there will be increases in PM and NO_2 concentrations in urban areas". But I think in their defence – and there's not a lot of defence, because there wasn't a lot of holistic thinking going on at the time – it really was about tackling climate change and CO_2. At the time the Euro standard – they were at Euro 3 then I think – was the big plan to increase emission control.' Yet the Euro 3 standard for diesel cars had a 500mg/km limit value for NOx compared to 150mg/km for petrol, and 50mg/km PM compared to just 5mg/km for petrol. If there was a conviction that the Euro standards would eventually save us, there was an equal acceptance that it was OK to expose the populace to increasingly poisoned air in the meantime.

In little over a decade, the number of diesel cars – all legally emitting several times more NOx and PM than the petrol cars they replaced – rose from under 2 million in the UK to over 12 million. James Thornton, the CEO of ClientEarth, a non-profit environmental law organisation, remembers appearing on a TV panel discussion when the VW story broke, alongside Sir David King, the government's former Chief Scientific Advisor from 2000 to 2007: 'He was saying, yes, we did believe diesel was lower CO_2 emissions … but we were hoodwinked by the car companies, we didn't realise the emissions were nearly as bad as they were or would be because the car companies promised us it was very easy to stick on filters, collect the particles and otherwise reduce dangerous emissions – and that didn't happen.' If the Euro standards were adhered to, 'they might have been a decent compromise', says Thornton. 'But they haven't been, so they were a terrible compromise.'

Diesel vehicles have consistently been found to fail the Euro emissions standards when tested in the real world. Researchers at King's College London undertook their own testing of over 80,000 vehicles at roadside locations in 2011 and found there had been little or no improvement in NOx emissions from diesel cars, vans, HGVs or buses for over 20 years (it did find a significant improvement for petrol cars), despite the Euro standards in place to reduce NOx. The study estimated that even the Euro 5 diesel cars, the best available at the time, emitted over 1.1g of NOx per km when on the road – more than five times the Euro 5 limit of 0.18g/km, and more than even the original Euro 1 'high bar' limit of 0.97g/km set in 1992. Other studies began appearing with similar findings. The European Joint Research Centre found that petrol cars largely performed within Euro emissions limits, but diesel cars emitted levels 4 to 7 times higher than they were supposed to. In 2010, when around 3.6 million Île-de-France (Greater Paris region) residents were exposed to NO_2 levels exceeding the annual limit, Airparif reported that, 'Although the filters that now equip most new diesel vehicles contribute to reducing particulate emissions, they also give rise to a significant increase in NO_2 emissions. It is now confirmed that the proportion of NO_2 in NOx emissions is increasing steadily.' In 2012, half of all private cars in London and virtually every bus, HGV, LGV and black cab, were running on diesel. That same year, just before London hosted the Olympic Games, the IARC upgraded diesel engine exhaust from 'probably' carcinogenic to humans to 'definitely'.

To make matters worse, the CO_2 advantage associated with diesels never materialised either. Research suggests that the global warming impact of increased black carbon emissions from diesel more than offsets the CO_2 saving, due to black carbon's ability to absorb and radiate heat. NOx also includes nitrous oxide (N_2O), which is a more potent greenhouse gas than CO_2, and diesel emits higher levels of NOx than petrol. Not just that, but many Euro 6 petrol cars now achieve almost the same fuel efficiency – and therefore CO_2 emissions per kilometre – as diesel. Europe gained all the ill-health for none of the climate benefits. And there were alternative options. To meet the same Kyoto targets, countries such as Japan, South Korea and the US chose to back research into low-emission hybrid and electric vehicles. Diesel in the US has always been 'socially and environmentally unacceptable', says Hochgreb. 'Why? Because European diesel fuel is a lot cleaner, much more controlled in terms of quality. The US fuel was not. It didn't work in the US because there were comparatively no regulations on fuel ... of course all of the freight uses diesel, terrible diesel, and terrible engines. But in Europe the incentives were there, there was a demand for higher efficiency than offered by petrol because of fuel prices and high taxes on fuel, much higher than the US ... The market share of diesel in Europe went to 50 per cent of cars because it was perceived as environmentally friendly.'

And only after all *that*, comes the VW scandal. The success of Volkswagen's diesel cars was based almost entirely on its environmental credentials. Riding high after the decade-long dash for diesel in Europe, VW tried to break the US market. It ran a commercial during the 2010 Super Bowl of its diesel Audi stopping by a long line of grey, wheezing, smoky cars, before being waved through by the 'green police' saying 'You're good to go.'

An in-depth description of the VW scandal in *Fortune*, March 2016, outlined Volkswagen's 'audacious' goal to become the biggest car company in the world. They saw cracking the US market as 'crucial to the mission'. However, California's emissions rules were getting in the way.[*] Chancellor Angela Merkel personally weighed in on the issue in April 2010 when she took on the head of the California Air Resources Board (CARB), Mary Nichols, in a private meeting. However, Merkel picked a fight with the wrong person. And given that the third person in the room was Arnold Schwarzenegger, that's really saying something. Mary Nichols, chair of CARB from 1979 to 1983, had returned for a further spell in charge – at the request of Governor Schwarzenegger – in 2003, and has stayed in that position ever since. She's rightly been described as a 'rock star' of the air quality world, and a known tough negotiator. When I spoke to her over the phone from New York, I asked if it was true that Angela Merkel asked her in that meeting to relax the NOx limits for the sake of the German auto industry? 'I actually had to testify on this under oath in the German Bundestag,' she laughs. 'She didn't *ask* me to do anything, what she *said* was – in front of my then boss, Governor Schwarzenegger – "Your diesel emissions standards are too severe and they are hurting our German companies". It was

---

[*] We tend to think of European environmental regulations being stricter than the US, but California's emissions regulations – for reasons I look at in detail in Chapter 7 – have consistently been a step ahead of the Euro standards.

an accusation, not a question, I guess in effect "you are doing something wrong, you should stop!"' And what did you say? I asked. 'I said "I don't think it's true. We need to have these controls because of the need to meet health standards." I answered back, being that sort of person! ... We knew already that the German companies were opposed to our NOx standards because we met with them all the time. But we weren't expecting to hear these things from the Chancellor.'

In Europe too, the Euro regulations were getting tighter. The PM emissions allowed for diesel passenger cars had gone down from 25mg/km in the Euro 3 regime to just 5mg/km in Euro 4 and 4.5mg/km in Euro 5, the same as petrol. The gap between diesel and petrol cars for NOx emissions was also closing. Whereas Euro 5 allowed diesel cars to emit up to 180mg/km of NOx and petrol cars 60mg/km, from September 2014 Euro 6 would allow just 80mg/km for diesel, while petrol would stay at 60mg/km. A 2014 EU report anticipating the changes concluded, painfully ironically in hindsight: 'The worry is that real-world emissions might not show the same decrease.'

Regulators have historically relied on highly controlled lab-based emissions tests at facilities like Millbrook. An automotive company brings their latest model in for testing, and the 'drive' takes place on rollers in the lab (known as 'dynos'), at an exact temperature and speed, ensuring that the test is precisely the same and repeatable for every model. You compare like for like, cancelling out all other variables. At Millbrook I got to watch one of these tests in action from the control room, viewing the car through a small, thick-glass window. A shiny new SUV is perched seemingly stationary while its wheels whirl around on the dyno below; a professional driver inside it slowly decelerates down from 80kmph to zero. In the next room a bag, like a giant sandwich zip bag, slowly fills with the exhaust fumes. I ask if I can touch it, and I feel the warmth of the smoke as it slowly inflates. The exhaust from the bag is then passed through various analysers to measure for levels of NOx, carbon monoxide and PM.

Unfortunately, as *Fortune* put it, 'that approach makes it possible to cheat'. Software now known as a 'defeat device' could be installed to recognise when the car was within lab conditions, and ensure the car only emitted what it was allowed to during the test and not a minute longer. The one installed by VW was found to increase NOx emissions to a factor of 10 to 40 times above EPA (Environmental Protection Agency) compliant levels as soon as it was out of lab test conditions.

In 2013, the US-based non-profit International Council on Clean Transportation (ICCT) began looking into the disparity between the claimed performance of European diesel cars and their real-world emissions, finding NOx emissions of up to 35 times permitted levels. In May 2014 it sent its report to the EPA. CARB and the EPA, both suspecting a defeat device, began scrutinising VW's diesel cars in minute detail for most of the following year. 'The ICCT organisation actually built the case, and brought it to us,' confirms Mary Nichols. 'The EPA joined with us in the prosecution of it.' Aware they were about to be found out, VW privately told the EPA on 3 September 2015 that their cars were indeed fitted with illegal software, perhaps hoping for a behind-the-scenes slap on the wrist. The EPA instead revealed VW's cheating via a public 'Notice of Violation' issued on 18 September 2015, stating unambiguously that 'VW manufactured and installed defeat devices in certain model year 2009 through 2015 diesel light-duty vehicles … These defeat devices bypass, defeat, or render inoperative elements of the vehicles' emission control system …' In the six-page public notice, the EPA also reiterated that the purpose of having emissions standards was 'to protect human health and the environment'. Within days, VW admitted it had installed defeat devices in up to 11 million cars worldwide, beginning in 2008.

'I was shocked by how long it had gone on,' admits Mary. 'Frankly, as an appointed official in the government I was concerned that we would be seen as, and perhaps we really were, at fault for not having found it sooner. I was angry about the cheating and wanted to make sure that we got it to stop, and punished the violations, but I was also worried that ... it might be a much more widespread problem.'

I asked Phil Stones at Millbrook how VW had managed to pull the wool over his eyes, and that of other testing facilities, for so long: 'The regulations are that what it does on the dyno it must do on the road, in the same conditions, or words to that effect. So, if you drive on the dyno, same conditions for me – this is my interpretation – is that it's that speed load, it's that time, it's that same window of operation.' The VW software could tell when the car was in test conditions because the car was running and the wheels were spinning, but the steering wheel wasn't moving: 'It doesn't see any steering input on the dyno, so it went to a mode that "defeated" the emissions.' VW cars were able to meet the Euro limits, but only for long enough to pass a test when no steering is registered on the steering wheel. On the road, as soon as the steering wheel moved, the VW could effectively loosen its belt and exhale high levels of pollutants once more. I asked Phil what the atmosphere was like in Millbrook the day after the scandal broke in September 2015. Millbrook staff work alongside engineers from all the major car companies every day – if they are not there for regulatory tests, then they hire out Millbrook facilities to test out prototypes. Although he's not allowed to confirm it, Phil most likely worked alongside VW engineers who knew their models were cheating, and watched as Phil and his team ran the tests. 'Everybody was surprised that they knowingly, directly used a defeat device,' says Phil. 'People "optimising" is different, it happens in any industry, sport, Formula One, anything – people optimise and make decisions commercially. But cheating puts them in a different world. It's like a sportsman – does he do high-altitude training? Or does he take steroids? That's the difference in my world. They took the steroid drug.'

Andy Eastlake, now managing director of the Low Carbon Vehicle Partnership (LowCVP), was a senior emissions engineer – and Phil Stone's former boss – at Millbrook, until

2011. 'For many years we were focused on air quality through the Euro standards, and all of the air quality modelling that we put in place assumed that we would deliver to those standards,' he tells me. 'I remember presenting on "real-world testing" papers back in 1997, 2000. Real-world testing is not a new issue. What has changed, and it was genuinely a shock to me, was the blatant fraud, if you like, against those regulations. Everyone knew that those regulations weren't necessarily as all-encompassing as they should be ... the risks of only testing a small area of operation and assuming it will be clean for the rest of the operation – the risk had gone up manifestly. And then you've got an individual company that has driven a coach and horses through that loophole.'

VW's was very, very far from being a victimless crime. National and city transport decisions are based on forecasted emissions data; the falsifying of that data led to many thousands of deaths. According to one analysis, the 11 million VW cars fitted with defeat devices would have collectively emitted close to 1 million tonnes more of nitrogen dioxide every year than the policy-makers, regulators or car owners were expecting. And according to the EEA (European Environment Agency), nitrogen dioxide pollution causes 78,000 premature deaths in Europe every year. Paul Bate was a senior transport engineer for Derby City Council from 2001 to 2007. 'I can remember the nice graph that said "this is the emissions from Euro 4, and this Euro 5, and by the time get to Euro 6 there is no air quality problem whatsoever". And what happened is we got to a certain point and then emissions flatlined ... vehicles were being optimised to pass a particular test, whereas in the real world they were performing sub-optimally. So what we ended up with is a disconnect in the information that planners in city authorities across the world have been given to plan for and improve air quality in their areas, it has been wrong by several orders of magnitude ... the standards, and all the tools and models, are based on the fact that vehicles perform according to the Euro standards. And there was a very real difference.'

In the weeks after the VW scandal, Ally Lewis and Frank Kelly penned a joint letter to *Nature*, stating that 'pollution from diesel vehicles has long been under-reported'. When I met Kelly I asked him whether they had suspected foul play. 'From the year 2000 we had predictions for what NO_2 would do in London, for example. And the predictions were all very favourable. There was a line going down. Our measurements, however, from 2000 to 2006–7, were a straight [horizontal] line. What we were predicting was going to happen to air quality was getting increasingly distant from what was actually happening. So, we started scratching our heads around 2005 to try to understand … We ended up getting in some laser-based equipment from America: you set it up at the side of the road, shine the laser across, and whenever a vehicle goes past it tells you what the NOx is in the vehicle exhaust plume.' After tens of thousands of different vehicles had passed the laser, Kelly's team found that what the vehicle was *meant* to be doing was nothing like what the vehicle *was* doing: 'We got the equipment from the university of Denver with funding from Defra [the UK Department for Environment, Food and Rural Affairs], and they got our report around 2007. And Defra basically sat on it. So, we did some more work organised by the manufacturer, showing that certain cars and vehicles were a lot worse and some a lot better than others, and Defra got that information as well. Clearly, they were uncomfortable about reporting this because it would bring certain car manufacturers into a bad light. We knew what the problem was, we had the data, and Defra had the data, long before – a couple of years before – the VW scandal broke in the States … it took the Americans to actually do something about it.'

There was, agrees James Thornton of ClientEarth, a 'grand collusion of most of the motor industry to delude the regulators and the public, although the regulators did know about quite a bit of it and did nothing'. He repeats Kelly's allegation that, 'in the UK the relevant agency did know and didn't do anything, it took the US Environmental

Protection Agency to blow the whistle ... the real problem has been the huge investment of a German car company in diesel engines and the control that the German car companies have over the German economy and therefore over a lot of the European economy. The reason Volkswagen diesel was eventually discovered was that Volkswagen had invested ... hundreds of millions in developing their new generation diesel engine and having spent so much on it they decided it was time for America to fall in love with diesel. So, they took it to the US and they neglected to understand they didn't have the same sweetheart deal with the US EPA, who actually tested it and discovered that they had these defeat devices.'*

By the time VW were at it, however, so were many of the other car companies. Phil Stones was happy to tell me about a few more: 'The Fiat one was a timer of 22 minutes. The test is for 20 minutes. Their defence was subsequently "well on the road, it does exactly what it does on the dyno". The dyno only goes for 20 minutes, so they defined that that's how long the regulators want it compliant for, and after that they can do what they like. So they did. They "modulated", as the Italian Transport Minister said, they modulated their emissions – their after-treatment – down, after 22 minutes ... Others put temperature boundaries in. The test was between 20 and 30°C – so on the road, between 20 and 30°C it did exactly what it did on the dyno. Drop to 17°C, it didn't do the same.' Because everyone was attempting to hoodwink the regulators, does it suggest that the Euro 5 or 6 regulations for diesel were becoming unachievable? 'Er, no. They were achievable,' says Phil. 'My view is that in a

---

* To further rub salt into the wound, it later transpired that VW developed the defeat device using a €400m loan from the European Investment Bank: the loan was supposed to help the carmaker develop an engine that would comply with increasingly stringent emissions standards. Which, I guess, they kind of did.

commercial world you will do as much as you need to do relative to your competitor.'*

An independent 'real-world emissions' league table, called the EQUA Air Quality Index, overseen by a number of academics from institutions including King's College and Cambridge, emerged following the VW scandal. It takes all new car models out on the road to see what really comes out of their exhaust pipes. At the time of writing there are eight models of diesel car manufactured to meet the latest Euro 6 standards that not only don't meet it, but don't meet any of the previous standards going back to 1993. Two Nissans, a Fiat, Subaru, Peugeot, Renault, Infiniti and Ssangyong (no, me neither), were all rated 'H', meaning 'No comparable Euro standard: roughly equal to polluting over 12+ times Euro 6 limit'. If you include models rated G 'No comparable Euro standard: roughly equal to 8–12 times Euro 6 limit' and F 'No comparable Euro standard: roughly equal to 6–8 times Euro 6 limit', of cars launched after the Euro 6 requirements of 2014, there are another 29 models, by manufacturers including BMW, Ford, Mercedes-Benz, Volvo and Vauxhall, to name just a few. If you add in all supposedly Euro 6-standard cars that only managed to reach Euro 3 rating (a standard that's fourteen years out of date), Euro 4 (nine years out of date) or Euro 5 (three years out of date) during EQUA's road test, then there are a total of 134 diesel car models, from virtually any car brand or vehicle size you could name: all met the official lab-based Euro 6 emissions test and were released onto the market, and yet all fall short when independently tested in real-world conditions. And how many Euro 6 diesels did actually manage to meet the Euro 6

---

* And it wasn't just the cars. A spot check of HGVs in the UK between August and November 2017 found that 7.8 per cent of the lorries carried some sort of emissions cheat device. Out of 4,709 lorries spot-checked, 327 had been doctored to switch off emission controls. If the same proportion holds true across the country then around 35,000 illegally polluting trucks are operating across the UK.

limits during the EQUA road test? Just ten (and, to give them some credit, six of them were made by Volkswagen).

We arguably have VW to thank for raising diesel emissions so high up the political agenda and into public consciousness. Ally Lewis in York even suggests that 'Dieselgate' was the year dot for everything that has happened to his field of research since: 'Knowing the rate of change that was occurring before VW and the rate of change that has occurred after VW … the rate has increased by a factor of 10,' he says. 'I suspect VW will go down in history as probably making the single biggest contribution to improvements in air quality in Europe … The 2040 ban on UK sales of petrol and diesel cars … that would never have happened without the VW scandal.'

Volkswagen's unintended altruism didn't come cheap, however. In the US it has pleaded guilty to three criminal felony counts and agreed to pay fines totalling $2.8 billion, plus a civil resolution of $1.5 billion. In December 2017, Oliver Schmidt, head of the company's environmental and engineering office in the US, was sentenced to seven years in prison by US courts. By February 2018, VW was reportedly facing a bill of $25 billion worldwide. In May 2018 Martin Winterkorn, the former CEO of Volkswagen at the time of Dieselgate, was charged by the US authorities with 'conspiracy to defraud' and a warrant was issued for his arrest. In June 2018, the chief executive of Volkswagen's Audi division, Rupert Stadler, was arrested in Germany as part of the emissions scandal investigation. However, because the light of scrutiny soon lit up other car makers, too, VW's overall market share barely wavered. And the scrutiny spread to encompass air pollution from other transport sectors too. It was the name 'Dieselgate' that stuck, not 'VWgate', and come 2018 it wasn't about VW any more, it was about the fuel itself: diesel.

★ ★ ★

Trains are undeniably the 'greener' option for travel compared to cars. Given that you can fit far more people into them, the

average fuel use per passenger will always be lower. But diesel
fumes from diesel trains expose passengers and freight workers
to a surprising amount of pollution. All commuter trains in
18 out of 26 transit agencies in Canada and the US are hauled
by diesel locomotives. About 20 per cent of Europe's current
rail traffic is towed by diesel, with the UK, Greece, Estonia,
Latvia and Lithuania around or above 50 per cent. Over half
of India's 20,000-plus trains also run on diesel, consuming
around 7.4 million litres of it a day.* A month-long emissions
study in Seattle found that the average PM2.5 concentration
was $6.8\mu g/m^3$ higher for people living near railway
lines. Inside the train itself, a Boston study in 2010 found the
mean PM2.5 reading to be $70\mu g/m^3$ in the front carriage and
$56\mu g/m^3$ in the rear (a separate study in Toronto also found
pollution levels to be 3.7 times higher in the front carriage
compared to the rear, so note to self: always sit in the rear of
a diesel-pulled train).

A 2015 survey of Birmingham New Street station, one of
the busiest in the UK, found hourly concentrations of up to
$58\mu g/m^3$ for PM2·5 and $29\mu g/m^3$ for black carbon. The
highest levels were associated with idling trains, with
concentrations up to six times higher compared to passing
trains. Having travelled to Birmingham New Street many
times in my life, it has a noticeably low ceiling height, giving
the feel of an underground station, with very little space for
the air to circulate. Unlike an underground system, where all
trains are electric, however, here the majority are diesel trains.
Yet we seem to have a blind spot for their emissions, compared
to diesel cars. If our everyday commute involved a giant taxi
rank in a huge underground tunnel, with all their engines
running, the problem would be immediately apparent. New
Street station apparently has an impulse fan system that
responds to the levels of carbon dioxide in the station: the

---

*Which is equivalent to the amount of beer that Germany drinks in
a month, according to 24coaches.com. Apropos of very little, but
come on, how could I leave out a stat like that?

fans begin to operate once carbon dioxide exceeds 1,000ppm and the speed of the fans increases relative to CO_2 levels. However, such fans *only* respond to carbon dioxide levels, the bare minimum to stop us from suffocating: NO_2 and PM levels could in theory continue to rise unabated.

When Cambridge University tested emissions at London's Paddington station in 2015, around 70 per cent of its trains were powered by diesel engines. It's a one-way terminus, meaning that trains must enter and exit from the same side, resulting in many 'idling' with their engines running rather than switching them off during turnarounds. Around 37 million passengers use it every year, many of whom use it twice every working day. The Cambridge team led by Dr Adam Boies found that Paddington station's hourly mean PM2.5 hit a maximum $68\mu g/m^3$, while NO_2 maxed out at 120ppb (which would be a breach of the EU hourly NO_2 limit of 105ppb, if the station didn't have a roof – but it does, so outdoor guidelines don't apply). Sulphur dioxide (SO_2) concentrations, virtually non-existent outside the station, averaged 25ppb. The peak concentrations between 7 a.m. and 10 a.m. corresponded precisely with peak idling train activity.

Despite my repeated interview requests over several months, Network Rail, the operator of train lines and stations in the UK including Paddington, declined to speak to me on the subject of air pollution and diesel. The same happened when I approached a number of private train operators directly, too. Despite being very happy to market themselves as the green way to travel,[*] they get very spiky when it comes to discussing diesel. Regarding the Paddington study, one harassed Network Rail press officer eventually responded by email: 'I am waiting for a report to be sent over to me by our environmental team which outlines the air pollution levels at

---

[*] Which I'm not refuting. One diesel train carrying 200 people is better by every measure than 200 people driving separate diesel cars. But 200 people in an electric train is better for everyone.

Paddington. Once I have this I will share with you but, to give you a flavour of what it contains, I am told that air pollution was not as bad as expected due to good levels of ventilation with the biggest pollution source was Burger King extraction fan.' When he sent me the report a few days later, its title – 'Air and health impacts of diesel emissions' – promise much. Published in January 2016 by the Rail Safety and Standards Board (RSSB), an 'independent' UK body with a membership comprised entirely of the UK rail companies, however, it is very obviously, and disappointingly, an industry whitewash. Its opening salvos talk up the role of diesel in UK rail 'for the foreseeable future' and play down the health concerns, mentioning just two studies where 'The carcinogenicity of diesel exhaust emissions appeared to be specific to rats … not relevant to humans' and 'no evidence of increased risk at lower cumulative exposure levels'. As we've seen in all the previous pages of this book, and will see even more evidence of in Chapter 6, that is total rubbish. Regarding Cambridge's Paddington study, the RSSB report is a virtuoso performance in deflection tactics: 'the position of the monitoring equipment suggests that the results may have been influenced by other sources of particulates independent from the railways. One of these was adjacent to the Burger King outlet … attributed to cooking emissions, corresponding to the busiest times for the outlet. The highest particulate concentrations were recorded on the Praed Street ramp, some distance from the platforms, where smokers habitually gather; it would therefore appear that these more local sources of pollution were significantly affecting the overall results.'

I sent the RSSB report to Dr Adam Boies at Cambridge. It was the first time he'd been made aware of it, despite it being out for over a year – clearly the RSSB had never contacted the academics involved to better understand their findings. Understandably, Boies was not impressed: 'We had set out to measure emissions directly from the trains, but were not allowed to do so by any of the operators,' he told me. 'Had we been able to do so, we would have been able to answer the question that is being pointed to by RSSB: were the high levels of pollutants in Paddington station a result of the cooking or the train emissions? At the time we conducted some "back-of-the-envelope" estimates of comparisons between likely emission rates between trains and the Burger King at Paddington and found it highly unlikely that the Burger King emissions were greater than the trains ... We did conduct chemical analysis of the collected soot samples to determine whether cholesterol (a marker for emissions from cooking) was higher in the particles near the station centre. The results of this analysis showed that there was not a measurable level of cholesterol in the soot samples near the station centre, thus the particles were not identified as being explicitly from cooking. The truth is that we could answer this question quantitatively if the diesel emissions were measured from the fleet of UK diesel trains. We were not allowed to do so.'

Frank Kelly at King's College London agrees that 'both NO_2 and PM2.5 are coming out of diesel engines in large emissions ... it's recognised that diesel trains need to be replaced ... At the moment there's no discussion around diesel trains affecting air quality and people's exposure – if we start having that discussion then you never know what might happen ... we need to replace diesel trains with electrified trains.' Instead, in July 2017, the UK Transport Secretary Chris Grayling cancelled the planned electrification of three major UK rail lines.

Another high diesel emitter that lurks under the radar of public consciousness is the diesel generator. Used to produce off-grid electricity, these range from small generator sets used by market stalls to large ones used on construction sites. The smallest class of diesel generator (below 19kW) comprised 18 per cent of the off-road machinery in the United States in 2004, but generated 44 per cent of diesel PM and 12 per cent of NOx emissions derived from mobile sources nationally. According to London's Atmospheric Emissions Inventory for 2010, non-road mobile machinery including generators accounted for 10 per cent of NOx in greater London and 11 per cent of PM10. An article in the April 2017 edition of

Environmental Scientist magazine headlined 'Will backup generators be the next "Dieselgate" for the UK?' asked why no public official log or database of diesel generators existed despite thousands running across the UK, arguing that the 'air quality impact from the operation of diesel generators has often been overlooked'.[3] According to the author, a relatively modest 8 megawatt (MW) 'genset' would emit NOx at the rate of 26.7 to 42.2 grams per second.

The use of diesel generators is greater still in developing and middle-income countries where the power supply can be unreliable. With not enough electricity to meet the urban demands of Nepal, one local newspaper reporter describes the familiar sounds that accompany blackouts: 'thousands of diesel generators rumble to life, spewing noxious particulate matter.' In Delhi, a local businesswoman who runs a factory with around 200 sewing machinists, told me: 'Of course there are diesel generators running everywhere … And why? Because our government cannot provide us with 24-hour electricity. You don't get enough power supply from the municipal supply, it is only that … they all have 24-hour diesel back-up, everybody has these massive generators. All construction uses diesel generators … You need an uninterrupted power supply. You can't hire labour and then ask them to sit down and get up, sit down and get up … It all boils down to the basics. Get those basics sorted out, and everything will get better.'

The sector perhaps most addicted to low-grade fuel, however, is shipping. It is easily the transport sector with the dirtiest history. Shipping emissions contribute nearly 15 per cent of NOx and 13 per cent of sulphur dioxide emissions globally, and these numbers are increasing. Due to growing populations and consumer spending, more and more supertankers set sail every year. Since 1985 global container shipping has increased by about 10 per cent annually, with only brief dips for each recession. The Third International Maritime Organization (IMO) Greenhouse Gas study forecasts that by 2050, CO_2 emissions from international shipping could grow by between 50 per cent and 250 per cent depending on future economic and energy growth. Crude oil accounts for roughly a quarter of all goods transported by sea. Which gives us the headache-inducing fact that a quarter of all shipping emissions come from shipping the fuel needed to produce the emissions.

In Europe, shipping in the Baltic Sea, the North Sea and the English Channel causes more than 800,000 tonnes of NOx each year (roughly four times the total amount that Belgium emits each year). Shipping fuel also contains up to 3,500 times the sulphur content of the diesel used by road transport. In southern California, says Sam Atwood at the South Coast Air Quality Management District, the twin ports of Los Angeles and Long Beach combined are the largest port complex in North America. 'Until recently the ships burnt some of the very dirtiest bunker fuel in the world,' he says, 'extremely high sulphur content fuel, and then once the ships get to the port the containers are offloaded onto diesel trucks, and then possibly taken to an intermodal centre where they are loaded onto diesel locomotives.'

After my bad experience with the railway authorities, I was pleasantly surprised when Simon Bennett, director of policy at the International Chamber of Shipping, agreed to talk to me. 'It's a complex issue,' he says when we begin. 'In shipping we're regulated by the International Maritime Organization, that's a UN agency, and pretty well all of our regulations are adopted using a global regulatory framework ... Shipping is inherently international, you're moving cargo from one country to another, so in simple terms if you had different rules at different ends of the voyage you'd have chaos ... for the best part of a hundred years we've actually had a framework of global regulations.' When it comes to environmental regulations, shipping is governed by the MARPOL convention, an international agreement originally established to address disasters such as oil spills, but recently expanded to encompass things like NOx and sulphur dioxide. 'The mid-1990s was the first time it actually set a cap for the sulphur content of marine fuel,' Bennett informs me, adding 'it was admittedly quite a high cap.' I am genuinely astonished by what he tells me next. 'For the last 30 years or so the majority of ships have actually been burning residual fuel, which is basically the dregs from the oil-refining process, like tarmac ... the reason why we've been using it is because the oil industry was very keen for the shipping industry to use residual fuel because in simple terms they had nothing else to do with it.' I ask if 'residual fuel' is also known as heavy fuel oil, or HFO? 'Yeah, HFO ... it suited everybody, it's provided the world economy with cheap goods transport, and as I say the oil industry were very keen to provide this, initially they were almost providing it free of charge ... Until about the late 1960s, I guess ships were actually using diesel and then they switched from diesel to residual because it was so much cheaper and ships were also becoming much, much larger as well.' Since then, millions of tonnes of HFO have been burned in densely populated port cities and the world's most pristine marine wildernesses. More than 850 ships operating in the Arctic today are thought to use HFO, according to the Clean Arctic Alliance, representing around three-quarters of total Arctic ship fuel use.

The leisure cruise ship industry, which bases its entire image on healthy, clean-living, outdoor experiences, is not much better. According to an annual survey of the biggest cruise ships operating in Europe, conducted by the German environment group NABU (Nature and Biodiversity Conservation Union), a mid-size cruise ship can burn through 150 tonnes of low-grade diesel each day, emitting as much PM as one million cars, as much NOx as 421,000 cars, and the sulphur emissions of 376 million cars (yes, you read that right – three hundred and seventy six *million* cars). The report authors accuse the industry of having 'contempt' for the health of its customers. I spoke with Sönke Diesener, transport policy officer at NABU, who told me, 'Shipping in international waters seems to be under the radar, nobody cares. Most ports are far from city centres ... Awareness is rising but unfortunately shipping is a very slow-changing business. The International Maritime Organization is dominated by pro-shipping countries, and small flag-of-convenience-states like Bahamas or Liberia (where ship-owners flag their vessels to save on tax and labour costs) have power to set the agenda.' According to NABU, just one large cruise ship in port will probably release more air pollutants than all the cars found within most port cities.

As with rail operators, I approached many ports and most didn't want to speak to me. Most have stuck their head in the (oil-slicked) sand on this issue, believing that ignorance is a preferable option. In 2017, a BBC investigation revealed that Southampton, the biggest cruise port in Britain, hosting some of the largest ships in the world, doesn't even monitor its air quality. Southampton City Council estimates that the port contributes up to 23 per cent of air pollution in the city, but the port either could not or would not verify this.

However, the Port of Venice's head of environment, Marta Citron, did agree to talk to me. Every summer Venice makes the headlines as local protesters campaign against the huge cruise ships that pull into port, overwhelming the small city with people and pollutants. Marta was accommodating and friendly, and passionately believes that cruise ships are a positive for Venice as a whole. 'It is the second-busiest port in the Mediterranean – the first is Barcelona,' she tells me. 'During 2016 we had more than 1,600,000 cruise passengers in Venice, and more or less 540 cruise ships ... a voluntary agreement taken by the cruise companies decided to use in Venice only ships smaller than 96,000 tonnage, since 2014 ... In 2007 the ships committed themselves to use 2.5 per cent sulphur content fuel from the entrance in the lagoon while at that time the MARPOL convention [required maximum content] was 4.5 per cent. Nowadays it is foreseen that all cruise companies that have signed the agreement commit themselves to use 0.1 per cent sulphur content in the fuel in Venice but during navigation they can still use 1.5 per cent ... we have calculated it could be as much as 90 per cent sulphur reduction.' I ask her about the protests of a few months previously, when nearly 20,000 Venetians voted in an unofficial referendum, with 99 per cent backing a motion to keep cruise ships away. 'Erm, we don't like to speak about this kind of things,' she laughs, nervously, 'it is a matter of communication more than environmental aspects, let's say. We are technicians ... the air quality is not a problem of Venice, it is a regional problem. In all the Venice region we have problems with PM, it is a continuous problem in relation to geographic and meteorological conditions.' She tells me that PM levels are higher in winter, when the cruise ships don't come, 'so it is clear that it could not be related to ships'.

When I put this point to Sönke Diesener at NABU, he is brusquely dismissive of this view: 'shipping contributes almost 100 per cent of the air pollution in Venice. The city has no cars and almost no industry ... the central mode of transport is shipping. But of course, you have to ask what kind of shipping and which kind of air pollution we are talking about. I guess the answer they gave to you was only based on international shipping and is maybe linked to PM10. For all other air pollutants the ratio is way higher ... We conducted measurements of ultrafine particles in Venice and the background level was around 5,000 to 10,000 particles per cubic centimetre. After arrival of a cruise ship it peaked at 360,000 and was on average above 60,000 ... The port city of Hamburg just published its most recent air quality report. In the city of almost two million residents, lots of cars and industry, ships account for 39 per cent of NOx emission. There is no doubt that this ratio must be much higher in a city like Venice.' As for higher levels in winter, PM10 is higher in most parts of Europe due to domestic wood fires and a lower atmospheric boundary layer. Northern Italy in winter is one of the PM10 hotspots of Europe. But nanoparticles and particle number are always higher close to the source – and in the case of port cities, that means shipping.

In many transport sectors, diesel is rapidly trying to clean up its act. Since September 2017, all new cars meeting the Euro 6 standard must also undertake a real world test on public roads – the emissions cap will be higher than in the lab tests, but it aims to bring them closer in line. In shipping, HFO will be banned by the International Maritime Organization from 2020, and fuel sulphur content will drop from 3.5 per cent to 0.5 per cent – albeit still a much poorer fuel than the diesel burned in road vehicles, which is typically lower than 0.005 per cent sulphur. But from the point of view of local air quality and nanoparticle emissions, there is simply no such thing as 'clean diesel'. Diesel is a fossil fuel with a very high mineral and carbon content: by definition, you can't burn it without health consequences for those that breathe it in. Forget about sulphur for a second – that is an issue that should have already been consigned to the history books, were it not for shipping literally burning tarmac in their engines. Even ultra-low-sulphur diesel fuel that meets the latest Euro 6 (or the equivalent US or Indian standards, EPA LEV III or Bharat 6) is terribly bad for us to breathe in. As Professor Paulson at UCLA describes, 'when you live close to a roadway you have higher exposures to ultrafines ... and you also have more directly emitted gases from all the little engines burning fuel ... ultrafines are uniquely able to go places in one's body that other particles cannot.'

Studies have found that nanoparticles smaller than 300nm (PM0.3) account for over 90 per cent of the total numbers of all particles. Remember the golf balls and the footballs? If you have the choice, it's better to breathe in a few PM10 footballs with lower surface area than thousands of nanoparticle golf balls with all that surface area to do toxic damage to your lungs and with the unique ability to get into your arteries. To avoid exposure to those, you need to avoid exhaust fumes. An experiment called DAPPLE (Dispersion of Air Pollution and Penetration into the Local Environment) highlighted a noticeable difference in the concentrations of nanoparticles between roadside and building-side locations in central London from 33,162 particles per cubic centimetre (cm^3) up to 163,110cm^3. Given that sides of buildings in central London are rarely very far away from a road, that's a big difference within a matter of a few metres. Professor Prashant Kumar, of the University of Surrey, found that the number of particles by roadsides (327,000 cm^3) in Delhi was ten times greater than background levels across the city (33,000cm^3)[4]. Other studies of major cities have found that the particle count decreases by up to 40 per cent of their kerbside level within a distance of 10 metres (33 feet), after which they bunch together to form footballs, or evaporate.

The drivers sitting in diesel vehicles are exposed to the worst doses of all. At the British Heart Foundation Centre of Research Excellence in Edinburgh, David Newby recalls that, 'The *Sunday Times* ran a mischievous article following one of my studies saying you shouldn't cycle in the city. In actual fact, if you take a particle monitor inside a car, the levels are often three times higher than outside. Often there are no cabin filters in the air circulation. The air inlets at the front of the car are usually taking in the exhausts of the car in front of you.' The closer you are to the road, he says, 'the more pollution you breathe', and the car driver is obviously the closest of all. As for cyclists and pedestrians, 'the difference is exponential; you only need to be removed by one or two metres. Proper segregated cycle lanes that are only a metre away from the road can be enough.' Jim Mills, with his experience servicing the AURN (Automatic Urban and Rural Network), backs this up: 'it's a well-known fact that if you're driving along a major road you will get a higher dose than a cyclist or a pedestrian most of the time. Because you're sitting in a bubble, you're not getting the benefit of the air moving around you and diluting whatever pollution is around, you are essentially breathing air which is the exhaust of the car in front. We did some work for *Dispatches* on [UK] Channel 4 where we measured inside vehicles, on a bike, on a pedestrian and in a bus, and the highest levels in every one of those studies that we did was the driver. So, the guys who are producing the pollution are actually getting the highest dose … most people believe when they go inside their car and shut the door that the pollution is outside and they're safe, but nothing could be further from the truth.' The average adult in the US spends 55 minutes each day driving or being driven inside one such bubble of pollution.

Yet, in the race to reduce PM2.5 and NOx emissions from diesel car engines, the nanoparticle count and NO_2 percentage may have gone in the other direction. At Millbrook, Phil

Stones tells me that as the after-treatment of car emissions has become more and more advanced, 'you start to create more primary NO_2. So in older vehicles, most of the NOx would have been NO. Now, most of it will be NO_2. In real terms, the NOx has gone down ... while the proportion of NO_2 may have been 5 per cent, it is now 50 per cent ... Total NOx has definitely gone down, but I don't think total NO_2 has.' One of the most common after-treatments is urea injection, which turns the NOx into ammonia and then, in theory, turns the ammonia into water and (inert) nitrogen N_2: 'after-burning treatment is really expensive – for heavy duty vehicles especially. It has quite a lot of intelligence to it – there are a lot of sensors that monitor and feed back, it doesn't just pump urea in because it feels like it. If you pump too much in, or when the exhaust isn't hot enough, you create ammonia [NH_3]. It's called 'ammonia slip' – there's a legal limit on the amount of ammonia you can slip. And of course you consume a lot, which has a disadvantage to the user because it costs money to put in, and inconvenience ... if its after-treatment isn't working right, it will give a lot of NOx output.' By comparison, when a petrol engine 'after treatment isn't working right, it will give out hydrocarbons and CO [carbon monoxide]. So therefore the cold implications from petrol aren't quite as bad in terms of NOx as a diesel.'

The particle filters fitted to modern diesel cars are also a double-edged sword. As the Euro standards for PM from diesel have gone down from 140mg/km in Euro 1 to 4.5mg/km in Euro 5 and 6, the only way to implement this was to fit filters that trap the particles. Particle traps are very effective in controlling the solid primary particles, including black carbon, with filtration efficiencies found to exceed 90 per cent. However, they don't affect the secondary aerosol particles that form in the air after emission. And they do nothing for ultrafine particles. In fact, they can create more of them.

To measure particle numbers at Millbrook, Phil and his team take a sample by 'attaching butanol to each particle and using a laser to count them', he tells me. 'That's anything from 23 nanometres upwards.' Modern gasoline direct injection [GDI] systems produce a high particulate number because they work at higher pressures. The Euro 6 emissions standards try to limit this by counting particles down to 23 nanometres, but this misses a lot of nanoparticles below the 23 nm cut-off point. 'While diesel engines produce a high particulate number due to diesel particulate filters, which essentially take the big particles and burn them off, but this potentially creates a lot of small ones. GDI can create potentially more than a diesel particulate filter vehicle as the pressure of combustion creates lots of small particles. People are looking at gasoline particulate filters to reduce this. Particulate number has tended to go up since diesel particulate filters came in as the particles got smaller, but more of them.' So far, he says, 'there has been no regulatory reduction in the amount of particulate number that a vehicle can emit, but I would guess this is likely for Euro 7 if I was a betting man, or they will increase the range down to 10-nanometre particles to capture more.'

Particle numbers are also higher in winter in cold countries, because more nanoparticles are formed at lower temperatures and are less likely to evaporate. In 2016, a work group report on ultrafine particles by the American Academy of Allergy, Asthma & Immunology reported that 'although improved engine and fuel technologies have significantly reduced the emission of particulate soot, nanoparticles can still be formed from vapor condensation and they can be even smaller than the emission particles'. Furthermore, the 'introduction of catalytic converters … had the unintended consequence of shifting the bulk of the particle size distribution of exhaust PM to smaller diameters of 20 to 30 nm. Particle mass decreases with catalytic conversion, but the number of particles in the nanoparticle range increases.'[5]

It has been a huge own goal. Because of, not despite, all the cost and sophistication being thrown at the modern combustion engine, it is creating more nanoparticles for us to breathe. And it is doing so on the roads where we live and work, between ground and head height. Trucks are even worse. In a roadway tunnel study, trucks and heavy goods vehicles were found to emit 24 times more fine particles, and 15–20 times the number of particles, per unit of fuel burned, compared to light duty vehicles. Numerous studies going back to the late twentieth century have now concluded that whenever you're near traffic, it's nanoparticles that dominate the air; with each metre you step away from the road, the particle number count declines.

It reminds me again of that line from Professor Peter Brimblecombe, that environmental crises often repeat themselves, caused by rapid population growth, urbanisation, and fuel shortages followed by a new fuel more polluting than its predecessor. Modern engine technology is solving the problems of PM2.5, just as 20 years ago they tackled PM10. But they have made ultrafine nanoparticles and the total number of particles even worse. And these are the ones that we now know can get into the bloodstream and cause the most health damage. At Millbrook, Phil tells me that some cars are getting to the stage that half the cost of the car covers the after-treatment systems fitted to meet emissions standards; and yet, they still aren't working to improve our health as they were designed to, and may in fact be doing the opposite (although this is not to say that old cars are somehow better – go back a decade or two and you may be breathing fewer nanoparticles, but with an added serving of SO_2, higher NOx, higher carbon monoxide and much more black carbon). It all starts to add up to the conclusion: does our entire approach of burning fossil fuels for transport, within the streets that we live in, need replacing?

CHAPTER SIX

Struggling to Breathe

'Every parent who comes to me now is talking about air pollution', Dr Ankit Parakh tells me in his consulting room. A pulmonologist in paediatric medicine at BLK Super Speciality Hospital, he is on the front line of Delhi's smog epidemic. The BLK is one of several huge, privately owned hospitals that sprang up since the Indian economy liberalised in 1991 and, like so much in Delhi, the divide between the haves and the have-nots is stark. Beggars, traders and rickshaws crowd around the hospital entrance and spill out into the road. Inside, the paediatric wing is full of kids with shiny Nike trainers and concerned parents tapping impatiently on smartphones. Framed photos of happy (and exclusively white) child models adorn the walls. Most patients are here for breathing and respiratory disorders. Approximately one in three adults in Delhi, and two in three children, have respiratory symptoms due to poor air quality.

'In India the main problem starts around September, October – it peaks in November, then there is a down phase from December to February, and again peaks around March, April and May,' says Dr Parakh. As we speak it is currently the November peak. 'Anywhere in these seasons sees a surge of asthma and wheezing patients.' Wheezing is a technical term. Asthma tends not to be diagnosed until around the age of five, because the symptoms in young children can resemble a wide range of other respiratory problems. Asthma-like symptoms therefore tend to be called 'pre-school wheeze', which sometimes develops into asthma, sometimes not. 'Children who are exposed to air pollution, even pregnant mothers … the risk of these children actually having wheezing or asthma is definitely there,' says Dr Parakh. 'It is also related to the birth weight of the child … underweight children will have more wheezing episodes, the episodes are more severe, they are more difficult to control. And now it has been shown air pollution is not just a trigger that increases an underlying asthma, it is also an inducer which actually can generate an asthma … And it is not just about respiratory issues, people have shown hypertension.' He pauses, and says again, eyeing the queue waiting outside, 'There is a lot of concern amongst parents.'

I don't keep Dr Parakh long, aware that his consulting hours are short. When I leave, however, he insists on showing me the way to the Metro station. I expect him to point out the direction as we reach the main BLK entrance, but in fact he walks me out and onto the road, into the throng of people and traffic that seemed so distant in the sterility of his consulting room. Crossing the road, his stethoscope and brilliant white coat wrap around him like a protective layer. No one dares bother him. A girl of maybe eight or nine is begging nearby. She is covered in dust, her hair matted to the thickness of a door mat. Every minute of her work and life takes place on Delhi's roads, breathing in some of the very worst air in the world. Dr Parakh shakes my hand and ushers me protectively towards the station's barriers, and I soon step into the clean carriages of a Metro train.

It's easy to take everything Dr Parakh said as a given, but the definitive link between modern air pollution and health has only been made relatively recently. It had long been suspected. In the 1950s epidemiologists in California such as John Goldsmith made the link between air pollution and heart attacks, but the causal link was elusive and other factors such as high smoking rates were hard to separate. The International Agency for Research on Cancer suspected that PM2.5 was carcinogenic to humans in 1988, but only definitively said so as recently as 2013. Professor Bert Brunekreef, chair of the European Respiratory Society Task Force on Air Pollution, admitted in 2016 that 'when I started my career in environmental health some 35 years ago, air pollution in western Europe was not seen as much of a public health problem.'

By the 2000s and 2010s, however, studies began coming thick and fast, showing that common pollutants in the air affect our health at every stage of life, in the womb, continuing through childhood, adolescence and adulthood into old age. Professor Brunekreef describes this as the 'life course' of air pollution on human health. And it is this 'life course' that I'm going to take you through in this chapter, from conception through to your likely premature death. Sorry, this might not be the cheeriest chapter to read. But boy, is it full of counter-arguments to the statement that 'air pollution never did me any harm'.

In fact, it affects us before we are even conceived. The incidence of infertility has been creeping up in industrial countries from 7–8 per cent in 1960 to 20–35 per cent by the mid-2010s. During those 50 years, sperm concentration (the amount of sperm per millilitre of semen) almost halved. The sperm count in American males is decreasing by 1.5 per cent each year, and you don't need to be a mathematician to work out that is unsustainable. Recent studies have started to suggest a strong link between ambient air pollution and this decline in fertility. Compounds of lead, cadmium, and mercury have long been known to damage the male reproductive system. Now smoke-derived PM2.5 and polycyclic aromatic hydrocarbons (PAHs) have been found to impair or disrupt sperm production too, even leading to DNA fragmentation in sperm. A long-term study in Taiwan of thousands of young adult males between 2001 and 2014, found that for every increment of $5\mu g/m^3$ of PM2.5 levels there was a decrease of 1.29 per cent in normal sperm morphology (the size and shape of sperm in a sample). An Italian study in 2003 found that tollgate workers – a cadre of society unusually exposed to traffic pollution – had significantly poorer sperm quality than other men from the same region.

If we do manage to conceive, then air pollution can damage the health of the foetus and lead to birth abnormalities. A study by Queen Mary University, London, in 2018, found that inhaled black carbon particles passed from pregnant women's lungs to eventually cause small black spots in the placenta. A study in Ohio in 2017 of women living with high PM2.5 levels in the month prior to conception showed they had a higher chance of having a baby with birth defects – the most common being cleft lip/palate or abdominal wall defects – compared to those that did not. In Wuhan, one of the most polluted cities in China, researchers looked at all 105,988 births delivered in that city between 2011 and 2013. When these births were overlaid with the carbon monoxide (CO), nitrogen dioxide (NO_2), sulphur dioxide (SO_2) and ozone

(O_3) readings, the risk of babies being born with congenital heart defects was higher among women with greater exposures. Previous studies in California and Australia also found a higher risk of artery and heart valve defects from increasing O_3 exposure during the second month of pregnancy.

High air pollution exposure also increases the likelihood of having a premature birth. A team of researchers from Stockholm, London and Colorado concluded in 2017 that as many as 3.4 million premature births across 183 countries could be associated with PM2.5, with sub-Saharan Africa, North Africa and South and East Asia most affected. India alone accounted for about one million avoidable premature births. A US study in 2015 found that just over 3 per cent (or 15,808) of pre-term births nationally per year could be attributed to PM2.5.

Smoking has long been known to produce underweight babies, so it's no big surprise that traffic smoke does much the same. A London study in 2017 found that among half a million newborns, high PM2.5 levels were associated with a 2–6 per cent increased risk of low birth weight.[*] Even in Sweden, with its comparatively clean Arctic air, newborns with relatively high exposure to traffic-derived NOx were consistently found to have smaller foetal growth in late pregnancy: for every $10\mu g/m^3$ increment of NOx, birth weight reduced by 9 grams. Most worryingly, a 2017 US study using six years' worth of data covering nearly a quarter of a million deliveries found ozone was associated with a significantly increased risk of stillbirth. Both long-term low-level and short-term high-level O_3 exposure consistently increased the stillbirth risk, leading the researchers to estimate that approximately 8,000 stillbirths per year in the US could be the result of O_3 exposure.

For those who have a successful birth, air pollution raises the risk of pneumonia among under-fives and lifelong lung conditions such as asthma. According to the WHO, 570,000 children under five die each year across the world from

---

[*] The reasons include placental inflammation, impaired oxygen transport across the placenta, unstable blood pressure and even inflammation of the foetal lungs.

respiratory infections such as pneumonia, while up to 14 per cent of children aged over five currently report asthma symptoms, almost half of them related to air pollution.[*]

The Columbia Center for Children's Environmental Health goes so far as to say that air pollution is the root cause of much of the ill health in childhood today. Young children breathe in more air than adults relative to body weight, meaning they are disproportionately affected by air pollutants compared to adults. Babies under the age of one tend to breathe 600 litres per kilo of body weight, per day. By the age of four, as we grow, it reduces to 450 litres; by age 12, it's 300 litres; and by the age of 24 it plateaus at 200 litres per kilo per day and stays there for the rest of adulthood. When exposed to a polluted environment, children therefore suffer the ill effects three times more acutely than adults. Babies' immune systems are also not fully developed and are more vulnerable to infections and almost defenceless against toxic exposure. Children are the first and worst victims of lead pollution too, because their body's immaturity makes them most susceptible to neurological injury, leading to lowered IQs, reading and learning disabilities, impaired hearing and behavioural problems including ADHD.

There remains, however, a question of causality around much health evidence regarding air pollution. By their very nature the studies are based on the epidemiology – i.e. health trends across a population. You can't put 100 kids in a lab, expose them to pollutants, then cut them open to see what happened. So epidemiological evidence will always suffer from the kind of argument that goes, 'just because 30 per cent of the population gets cancer and 30 per cent of the population eats cornflakes for breakfast, does not mean that cornflakes cause cancer'. When an epidemiology study is repeated in different places, and consistently comes up with the same results, however, it becomes very hard to ignore. And the epidemiological evidence for air pollution reducing the size of children's lungs is, I would argue, the most compelling example.

---

[*] The WHO also suggests that warming temperatures and ever-rising carbon dioxide levels may increase pollen levels, making asthma rates even worse.

The Californian Children's Health Study is one of the most comprehensive investigations of the long-term consequences of air pollution that we have. Starting in 1993, more than 11,000 schoolchildren were selected from 16 communities, and their lung function measured annually while air pollution levels were measured continuously. The performance of the kids in forced expiratory volume or FEV tests (how much air a person can force out in one second) declined according to exposure to NO_2 and PM2.5. The proportion of 18-year-olds with a low FEV was four times higher in the communities with the highest PM2.5 levels compared to the lowest. Their lung growth had actually been stunted. Living within half a kilometre of a freeway was associated with a 2 per cent deficit in forced vital capacity (the total amount of air in the lungs). International study after international study has since found the same, including in Mexico, Austria, Norway, Sweden, the UK, and the European Study Cohorts for Air Pollution Effects. A three-year study in China found that an increase of $10\mu g/m^3$ of PM2.5 was associated with a loss of 3.5ml in FEV capacity. In Delhi in 2012, one in three children in the city were found to have reduced lung function.

Professor Chris Griffiths, from the Centre for Primary Care and Public Health at St Barts Hospital and a working GP, was involved in a six-year children's lung capacity study in London which concluded in the 2010s. The children living in areas with high levels of particulates and nitrogen dioxide had their lung capacity reduced by up to 10 per cent. I asked him if there remained a 'causality problem': 'You're not going to get a randomised clinical trial of air quality interventions, it doesn't work like that,' he argues. 'You've got to reach a point where you say "Well how strong is the evidence? What's the quality? What's the causal inference?"' Regarding his own lung study, he says 'the mechanisms that underlie these observations are not clear, but that doesn't mean that those associations aren't there or aren't important, just because we don't quite know how air quality mechanistically reduces lung growth. But because the data is quite consistent wherever these studies have been done, Europe, Scandinavia, Boston, California and then most recently London, studies in different settings with different pollutants, yet they appear to be saying similar things about lung growth.' The conclusion drawn by many, including the UK's Royal College of Physicians, is that there is little reason to 'doubt that air pollution adversely affects the normal growth of lung function during childhood, right up to the late teens'.

The ongoing Californian Children's Health Study, led by William 'Jim' Gauderman, a professor of preventive medicine at Keck School of Medicine USC, also continues to add to the weight of evidence. While in 1993 Los Angeles had some of the worst atmospheric pollution in the world, by the 2010s it was still bad but much improved. His recent papers have been able to flip the arguments against epidemiology on their head. Cornflakes don't cause cancer. But if you reduce the consumption of cornflakes* and the prevalence of cancer drops at precisely the same rate, then perhaps you *do* have to question what you're putting in your bowl each morning. Comparing three cohorts of the Californian Children's Health Study from 1994 to 2010, Gauderman and his team found that the mean four-year growth in FEV (volume of breath in one second) increased by 91ml for every decrease of 14ppb in NO_2, and similarly for PM2.5:[1] in other words, whenever the nitrogen dioxide and PM2.5 levels went down, the children's lung function showed a marked improvement.

As our life course of air pollution moves from childhood and into adolescence, low FEV comes with an increased risk of cardiovascular disease – given that you're missing 20 per cent of your possible lung capacity, you're less able to exert yourself during exercise. Equally logical is the fact that asthma is both caused by high pollution and irritated by high pollution: ozone, nitrogen dioxide and PM2.5 all cause inflammation of the airways, and airway hyper-sensitivity is the defining characteristic of asthma. The UK Chief Medical Officer's report suggests a link between diesel particles and the high asthma rate of 1 in 12 adults and 1 in 11 children in the UK. But respiratory conditions are only the most obvious ones

---

* Other breakfast cereal analogies are available.

affecting young adults. Our brain capacity may also be reduced upon entering adulthood. European studies have associated traffic pollution with lower cognitive development in junior school children followed by sustained attention deficit in adolescents. Studies in Mexico City have also revealed elevated levels of inflammation in the brains of children exposed to high air pollution, resulting in cognitive deficits.

If you grew up in a rural oasis of clean air and avoided all the health problems thus far, then if you move to a city (or your village *turns into* a city, as several have in Asia in recent decades) at any point in adulthood, plenty of health problems still await. Sticking with the brain for the moment – and let's face it, brain damage is one of our biggest fears – a Chinese study exposed lab mice to NO_2 inhalation and found it caused deterioration of spatial learning and memory. The study authors memorably describe air pollution as 'a multifaceted toxic chemical mixture capable of assaulting the central nervous system'. Outside of the mice fraternity, numerous epidemiological studies have linked NO_2 pollution to an increased human risk of neurological disorders, including reduced cognitive and attention scores. One fascinatingly precise study of white matter loss among elderly women, made using MRI (magnetic resonance imaging) scans, suggested that for every $3\mu g/m^3$ increase in PM2.5, white matter loss increased by 1 per cent. Suicidal depression has even been suggested by more than one study to increase one to three days after a peak in PM2.5 and NO_2.

For those who trust their gut more than their brain, a team from UCLA found that exposure to air pollution changes the composition of our gut bacteria. This leads to a whole range of health problems, including the circulation and build-up of cholesterol in the bloodstream. Other studies have found an association between air pollution and intestinal disease, appendicitis and even digestive tract cancers. There's a direct link for the chunkier PM10 here too: while large enough to be ejected from our throat and lungs via mucociliary clearance (our respiratory system's first line of defence, whereby a layer of fluid and mucus is constantly being propelled upwards for us to spit out), anything on the surface of the PM10 can be dissolved by saliva and find its way down into our gut. Depending on the chemical nasties on the surface of these coarse particles, this can lead to an imbalance in gut bacteria, or cause the chronic inflammation that leads to appendicitis or cancer.

Many airborne pollutants are known carcinogens. Polycyclic aromatic hydrocarbons (PAHs), for example, have toxic effects which can cause cell damage, leading to mutations and tumours. Long-term occupational studies of workers exposed to PAHs have shown an increased risk of skin, lung, bladder and gastrointestinal cancers. The PAH known as benzo(a)-pyrene, emitted in high abundance by stubble burning, was named as a human carcinogen as early as the 1980s by both the International Agency for Research on Cancer (IARC) and the US EPA. Since then the EPA has classified other PAH compounds as carcinogenic, all with confusingly complex names such as benz(a)anthracene, benzo(b)fluoranthene and indeno(1,2,3-cd)pyrene.

In summary, pick any major organ or body part and there will be a disease or defect related to air pollution. How about breast cancer? In Hong Kong, a $10ug/m^3$ increase annual PM2.5 exposure was found to increase the risk of breast cancer by an astonishing 80 per cent. Did someone say kidneys? Research from St Louis, Missouri, looked at more than eight years of data from nearly 2.5 million military veterans, and found that the veterans' kidney function worsened over time according to the level of pollution they were exposed to. Higher PM concentrations were associated with an increased risk of end-stage renal disease, after which a person requires kidney dialysis to stay alive.

Air pollution can even change how our DNA behaves. Genes – the segment of DNA that tell the cells of the body what to do, and when – are controlled by a chemical switch known as a methyl group. These methyl groups can, in effect, switch a gene on or off. A 2014 study by the University of British Columbia put 16 volunteers into an enclosed booth for two hours, giving half the participants clean air to breathe while the other half breathed diesel fumes equivalent to a busy highway. The methyl groups changed at about 2,800 different points on the DNA of people who breathed in diesel fumes, affecting about 400 genes. No similar changes were seen among the group breathing clean air. Until that experiment, scientists mostly thought that DNA responded primarily to long-term exposures. A similar Chinese study in 2017 compared traffic police officers to office-based police officers and found that DNA damage was significantly increased among the traffic cops compared to the pen-pushers at City Hall.

But it's the effect on our cardiovascular system that is most fatal across the adult population. Yes, even more so than cancer or lung disease. Air pollution causes thinning arteries, blood clots, heart attacks and strokes. As fine nanoparticles enter the bloodstream through the walls of the lungs, they cause increased inflammation, resulting in changes in heart rate, heart rhythm and blood pressure. This is not just from chronic, long-term exposure, but also from short-term. In Beijing, data from daily cardiovascular emergency room visits were collected from ten large hospitals for the whole of 2013, the year of the now-infamous Airpocalypse. A $10\mu g/m^3$ increase in PM2.5 was associated with a 0.14 per cent increase in cardiovascular emergencies per day. That may not sound like much, but given the monthly high and low of PM2.5$\mu g/m^3$ in Beijing can differ by as much as $300\mu g/m^3$, that could cause an increase in cardiovascular emergencies of 4 per cent, a significant burden on health services.

Years before David Newby's gold nanoparticles study (see Chapter 3), his first exposure chamber study exposed volunteers to street levels of air pollution and found that blood vessels were more likely to clot compared to being exposed to clean air. 'I like to call it a vascular stress test,' he says, 'What we do is put a little needle in the artery in the arm, and through that we infuse some agents to stimulate the blood vessels to relax and dilate. What we were able to show is when you are exposed to dilute [vehicular] exhaust, your blood vessels don't relax as much … and that means blood flow might be slower. We also tested a protein that is released from some of these cells, called TPA, which stops clots forming within the blood vessel, so the blood continues to flow … It is a clever way the body is able to both continue blood flow but also prevent you from bleeding to death. What we found is the release of this

TPA is lower when exposed to air pollution than if not. So, the defence mechanisms are hindered.'

Newby's team took this research a stage further, taking blood from a human volunteer and passing it through an artificial coronary artery. Inside this 'artery' was a strip of pig aorta – one of the heart's main pumps, at the top of the left ventricle – obtained from an abattoir. By cutting some of the surface off the pig aorta, it modelled what might happen when you have a heart attack and part of the artery bursts, exposing the deep layers of the artery.* 'What we found was that when you expose people to dilute diesel exhaust, the amount of clot formed on that strip increased. And when we put a filter in and took the particles out and redid the test again, the amount of clot came down to normal levels. So, it does seem that the diesel exhaust exposure makes the blood thicker. So that's three effects we've got: one, the blood vessels don't relax as much; second, they don't release as much of this clot-dissolving protein TPA; and when an artery is damaged, more clot forms. These are powerful mechanisms in the relationship between heart attacks and strokes.' I ask if he also attempted this test with gas pollutants, without PM. 'That's right, we did some NO_2 exposures, and also did it with ozone – just the gases on their own, no combustion-derived particles, and we didn't see any effects. People have argued with us about that, they were expecting certainly NO_2 to

---

* Of all the scientists I spoke to, or read about, for this book, I felt that David Newby pushed the boundaries more than most in the quest for the causal link to air pollution. I was surprised when, in early 2018, a scandal broke in Germany following the revelation that VW had taken part in an NO_2 exposure test on 25 human volunteers. It led news bulletins not just nationally but internationally. Chancellor Angela Merkel called it 'wrong and no way ethically justified'. Newby had been conducting similar studies for years. The German newspaper *Tageszeitung* probably put it best in saying that 'human volunteers only had to inhale exhaust fumes for a few hours, people in cities … have been breathing in levels of nitrogen oxide far higher than EU limits for years'. Quite.

have an effect, but we didn't see anything. Some people have argued that it is the combination of the NO_2 with the particles that causes a problem, and that is possible.' However, a team from University Hospital Jena, Germany, subsequently found a direct link with NO_2 too. Their 2018 study of 693 heart attack patients found that a rise in NO_2 of more than 20µg/m^3 within 24 hours increased the risk of heart attack by up to 121 per cent, while a rapid hourly increase of NO_2 of just 8µg/m^3 increased the risk of heart attack by 73 per cent.

If we actually manage to reach old age, then air pollution severely undermines our quality of life. In a study of the elderly in the United States, PM2.5 and NO_2 exposures were significantly associated with type 2 diabetes. Associations between air pollution and serum glucose, a measure used to assess diabetes status, have been reported even with short-term NO_2 and PM2.5 exposure. The same mechanisms that affect the young developing brain also set to work on diminishing brain function towards the end of life. Experimental studies have shown that air pollutants cause neuro-inflammation, neuron damage and blood–brain barrier problems. One South Korean study of Parkinson's disease hospital admissions from 2002 to 2013 found a 'consistent and significant association' between short-term exposure to air pollution (except for O_3) and higher rates of Parkinson's disease patients being admitted to hospital. The US Women's Health Initiative Memory Study (WHIMS) also enrolled over a thousand older women with no previous signs of dementia between 1996 and 1998, and studied them over six to seven years with regular brain MRI scans. The women who lived in places with higher long-term levels of PM2.5 were found to have smaller brain volumes, not explained by demographic factors, socioeconomic status, lifestyle or other health characteristics. Frank Kelly's research group at King's even looked at 8 years worth of GP records of over 100,000 Londoners aged 50–79, and found that those living in areas with high NO_2 and PM2.5 had a 40 per cent greater risk of developing dementia than those living with low pollution. The potential mechanism for this was suggested by work based in Lancaster, England (a relatively clean-air city, by UK standards). It found that local traffic pollution included 200 million metal nanoparticles per cubic metre, which it believed caused inflammation in the brain. I talked to Jim Mills, MD of Air Monitors, shortly after this study was released. 'Can you imagine what the political change of opinion would be if that gets proven?' he asked. 'That all the problems we have with rising dementia in our population, which is one of the most scary things I think that we face at the moment, are actually down to our use or overuse of the internal combustion engine producing these tiny particles?'

Underlying this dementia research, and almost all the health problems throughout the life course of air pollution, is oxidative stress. Oxidation is happening all the time, within our bodies and throughout the natural world. Highly reactive compounds search for electrons to steal from others, which sets off a chain reaction as compounds either lose an electron or try to replace one. The easiest example to picture is rust. When iron meets oxygen it creates iron oxide, or rust – oxygen steals electrons from iron, which forms a new, weaker compound. This flow of electrons is necessary for life to exist, driving photosynthesis by oxidising water (H_2O), which releases electrons to turn carbon dioxide (CO_2) into carbohydrates and oxygen (O_2). It's literally why we breathe: oxygen comes in and reacts with big hydrocarbon molecules like sugar to break them down and give us energy. Almost everything eventually oxidises down to CO_2 or H_2O, which we breathe out and pee out, and the process continues *ad infinitum*. So, oxidation equals good, right? Not completely. Oxidation is both caused by and causes free radicals, such as hydroxyl (OH), those atmospheric firecrackers that for the milliseconds of their existence are desperately looking for a fight. Again, free radicals are natural and necessary, and form part of our immune system. It's just that we don't want to encounter an unnatural number of them. If we do, it causes more oxidation than our bodies can deal with: this is known as oxidative stress.

Professor Frank Kelly, chair of the Committee on the Medical Effects of Air Pollutants, took up his first lectureship in the late 1980s looking at the effect of the free radical OH on premature babies. 'Because their lungs are under-developed they have to get extra oxygen in incubators to keep them alive and their brains functioning. But in giving their bodies more than the 21 per cent oxygen which we breathe normally – some of these kids require 90–100 per cent oxygen – it actually ends up damaging the tissue and they have a whole range of conditions that develop in their eyes, brains and lungs, because of this high oxygen concentration. And the reason this happens is because of free radicals.' Kelly's work in this then brand-new field of research found that while the lung tissue is protected from free radicals by the natural antioxidants in the body, if OH enters in high enough concentrations and for long enough, then the natural defences are overwhelmed – the atmospheric guard dog is unleashed, if you will, desperately snapping at soft body tissue and causing inflammation.[*]

Our body is under constant oxidative stress, even without pollutants getting involved. It causes damage to cells, proteins, DNA, and may even be the whole reason why we age. 'It's a bit like if you leave butter out too long it goes rancid and oxidises,' says David Newby. 'There is a theory that atherosclerosis, this fatty deposit in your arteries and the underlying cause of heart attacks and strokes … gets oxidised by pollutants, and it's this oxidisation that causes heart disease.' The bad news is that NO_2, pumped into every town and city in the world by car engines and boilers, happens to be a highly reactive free radical. Ozone, while not a free radical, is highly reactive and has an abundance of oxygen atoms, so tends to get into a fight with free radicals wherever it appears. PM and black carbon are coated in toxic material full of potential free radical stimulants. Our bodies' defences are therefore overwhelmed by this multi-pronged attack. 'Certainly when you look at the oxidative potential of air pollution particles, you really get a huge signal,' says Newby. 'In one of the studies I was telling you about – blood vessels not relaxing as much – part of the reason

---

[*] Many animals are affected in the same way, too. Studies have found that European house sparrows, blue tits and great tits from areas with higher vehicle pollution had increased levels of oxidative stress compared to birds in less polluted areas, while street dogs culled in Mexico City were found to have far greater lung and brain inflammation than rural dogs.

that we think they don't relax is that the oxidative stress consumes the mediator which makes the arteries relax … before it can have an effect on the blood vessels.'

In 2017, another team from Edinburgh Napier University looked at the release of defence proteins and peptides, which effectively march out to defend the body at the first sign of trouble. One such peptide, LL-37, which is present in saliva, tears and lung fluids, has many immune system functions including guiding inflammatory cells to a wound or infection. The Napier team examined the effect of black carbon particles of 14nm (within the range of particles able to enter the bloodstream) on LL-37. Even with relatively low concentrations of nanoparticles, the presence of LL-37 seemed to reduce; at high concentrations, it disappeared. On further investigation it was found that the black carbon particles had grown in size. Like a cartoon snowball rolling down a hill, they had simply stuck the peptides onto their own surface, and in doing so rendered them powerless. The bundles of carbon particles and LL-37 no longer had any effect on the bacteria present in the test.[2]

To make matters worse, according to Frank Kelly, activated inflammatory cells also generate and release large quantities of free radicals themselves as a form of defence. In the absence of any invading organisms to kill, these free radicals can turn on their host and start attacking local cell tissue components. Many of these reactions happen first in the lung lining fluid, the first line of defence the pollutants encounter when we breathe them in. Chemical chaos ensues, and an inundated immune system resorts to calling the body's last line of defence: a tank battalion of inflammatory cells. That chain of events precedes anything from an asthma attack to the early formation of tumours. Presumably, then, taking antioxidant supplements could counter this effect? 'If you look at trials that give antioxidant vitamins, they have absolutely no benefit whatsoever,' says David Newby, disappointingly. 'It doesn't matter how many antioxidants are flowing around in your bloodstream … that local burst is very difficult to prevent.'

The heart is particularly susceptible to oxidative stress, as it is an extremely active organ with a high metabolic rate and high energy demand. It's not exposed to the air, so can't oxidise and spoil in the same way as a lump of butter, but nanoparticles entering into the bloodstream can carry oxidative molecules on their surface. According to Newby's gold study, once inhaled, nanoparticles of 30 nanometres (83.3 times smaller than PM2.5, most coming from car engines) can travel anywhere the blood goes, which is basically our entire body. They are reactive and toxic due to their large surface areas (remember it's golf balls versus the footballs?), leading to oxidative stress and inflammation. The European 'Exposure and Risk Assessment for Fine and Ultrafine Particles in Ambient Air' study investigated the health effects of nanoparticles in coronary heart disease patients. It observed that the risk of developing ischemia – when blood flow to your heart is reduced, preventing it from receiving enough oxygen – was significantly greater two days after exposure to increased outdoor levels of nanoparticles. A US report on nanoparticles similarly concluded that they have 'greater oxidant potential and were much more prone to introducing cellular injury compared with PM10 and PM2.5'.

Ken Donaldson, a recently retired particle toxicologist from the University of Edinburgh's MRC Centre for Inflammation Research, spent the latter half of his career studying the toxicological impact of nanoparticles. I ask him if, in general, nanoparticles from combustion sources are more toxic than those from non-combustion sources. 'Yes', is the simple answer: 'combustion-derived particles are the big hazard because they have more than just a larger surface area, they have metals and organics. Both of these can undergo redox cycling[*] to produce oxidative stress.'

---

[*] Redox is a portmanteau word combining 'reduction' and 'oxidation'. As one molecule loses an electron (thanks to a scrap with a free radical) it is oxidised, while the molecule gaining an electron is reduced. I know, it sounds like it's the wrong way round. Thanks scientists. The other way to think of it is the oxidation state of a molecule is either increased (oxidised) or decreased (reduction) – and so on, in an endless cycle.

Add all these health effects up, and what do you get? According to the WHO, the answer is 4.2 million premature deaths a year (or 7.4 per cent of *all* deaths) from outdoor air pollution. The WHO even provides exact figures for each country. In the Philippines in 2012, for example, the WHO attributes 28,696 deaths to ambient air pollution. But unlike during the great smogs of London and Donora, these aren't numbers that represent people suddenly dropping down dead from air pollution. So how do we end up with such precise figures?

Frank Kelly, chair of COMEAP (Committee on the Medical Effects of Air Pollutants), tells me these figures are calculations based on the total years of life lost: in the UK, air pollution is shortening life expectancies by three to seven months on average, amounting to 340,000 years lost across the total population. Divide this by the average lifespan, and you get to a figure of around 40,000 'deaths'. Kelly even uses the word 'guesstimation'. Given that almost every article on air pollution leads with these annual death figures, isn't that a bit problematic? Numbers of deaths, says Kelly, is something that the public can easily understand, whereas 'they don't understand and don't worry about losing three months off our lives. But even that is a generalisation – some people are losing a day and some are losing ten years ... My response is, we talk about 89,000 premature deaths from smoking in the UK each year – Department of Health, official figures – that uses the same terminology, the same approach. It just helps to put it in terms of the risks we face in society. Breathing poor air in cities kills more people than drinking, more than obesity, way more than road accidents. But don't take the numbers of deaths as being 100 per cent accurate, because we haven't got that precision.'

COMEAP has regularly stressed that air pollution is shortening the lives of many more people than just those 40,000 a year in the UK. We will all die. But air pollution is likely to make you and me die sooner than we would have done otherwise. And, as it is related to many chronic, debilitating diseases, it is likely to make some of the years that we live much more painful than they would have been otherwise. The UK Chief Medical Officer's annual report recommends 'not just focusing on mortality, but using data that capture the full health consequence of pollution on morbidity, mental health impacts, and impacts on quality of life … Life-years (or quality-adjusted life years) are more appropriate for analysing policies than numbers of deaths, as it is when people die rather than whether they die that matters.'

Professor Chris Griffiths at St Barts Hospital believes the annual death figures aren't entirely helpful: 'there are other important ways of expressing the adverse health effects … I think the lung growth data is a little bit easier for people to get their head around, by saying "These children's lungs aren't as big as they should be because of air pollution".' I ask him if the focus should shift to quality of life. 'Yes, yes. You've got more people with asthma, people with asthma having more asthma attacks, more pneumonia, more hospital admissions, more strokes, heart attacks, premature births, small birthweights, these are all statistically significant adverse health effects and they all add up, as you say, to impaired quality of life. I think we are getting stuck on the issue of mortality and air quality … the truth is if people's lives leading up to that point of death are miserable because their lungs are destroyed … then air quality is important.'

Our life expectancy, then, is being reduced on average in Europe by 8.6 months to a year by PM2.5 pollution and in India by 1.1 to 3.4 years (or as much as 6.3 years in Delhi). But take away the source of the PM, and we see health dramatically improve. Frank Kelly writes of 'consistent evidence that a reduction in the level of particulate pollution following a sustained intervention (mainly regulatory actions) is associated with improvements in public health'. The WHO argues that 'health improvements can be expected to start almost immediately after a reduction in air pollution'. The deaths of many people every year are undoubtedly being caused by air pollution, although we'll never have an exact number. But all of our lives are being shortened, or worsened, by air pollution. If we want it to stop, we have to fight back.

PART TWO

FIGHTBACK

CHAPTER SEVEN

The Greatest Smog Solution?

London: 2015–2020

London is slowly waking from its smog-induced slumber and fighting back. The first day after the Labour Party candidate Sadiq Khan won a famous mayoral election victory in May 2016, he announced plans for a London Ultra Low Emission Zone (ULEZ), effectively taxing heavily polluting vehicles out of the city centre. Choosing a school in east London to make the announcement, Khan stood in the playground and told the assembled media, 'I have been elected with a clear mandate to clean up London's air – our biggest environmental challenge … I want to act before there's an emergency.' Just two days later, the new mayor found an old report buried in predecessor Boris Johnson's files. Left unpublished, the report, titled 'Analysing Air Pollution Exposure in London', had found that 433 of the city's 1,777 primary schools were in areas where pollution breached the EU limits for NOx. Of those, 83 per cent were considered deprived schools, with more than 40 per cent of pupils on free school meals. The poor were the hardest hit by air pollution. The news quickly appeared on front pages and homepages.

In the months that followed, Khan introduced a new air pollution alert system, giving pollution warnings on electronic signs at bus stops, tube stations and at the roadside, as well as issuing social media and text alerts. On 23 January 2017, the mayor tweeted: 'The shameful state of London's toxic air today has triggered a "very high" air pollution alert'. People with children aged between three months and five years were advised to keep them inside as a result of the widespread toxic air. He set a target of only buying electric or hydrogen buses from 2020, and introduced a £10 charge for older, more polluting cars to drive into central London. He also repeated the air pollution schools study that his predecessor had tried to shelve. Now, rather than 433, the number of affected schools had passed the 800 mark – all exposed to levels of NO_2 that breached EU legal limits. It was splashed across newspaper front pages that weekend, and the usual cries of 'what can be done?' were given a clear answer by Khan himself: a national diesel scrappage scheme, and a new Clean Air Act.

Leonie Cooper of the London Assembly, a long-time friend and colleague of Khan's, tells me that the Mayor takes air quality personally: 'He ran the London Marathon, and did a lot of training near roads – and I think, and this is my guess, that's when he developed adult-onset asthma. This is a very personal campaign for him. It is about health … as part of his campaign even before he was elected, Sadiq did a whole day with a King's College air pollution monitor and went to Putney High Street with me, and then on to Oxford Street, and to the North Circular ring road … it was pretty bad … in March [2017], he put the last of the zero emissions buses onto Putney High Street.' By March 2018, levels of NO_2 in Putney High Street were up to 90 per cent lower. Modelling of the ULEZ that now spans central London suggests it could lead to a 51 per cent reduction in NOx emissions, a 64 per cent reduction in PM emissions and a 15 per cent reduction in CO_2 emissions by 2020. By 2020 all single-decker buses operating in central London will be fully electric or hydrogen. In 2018, the Mayor stated his ambition for 80 per cent of all journeys in the city to be taken on foot, by bike or by public transport by 2041.

In July 2017, Khan found he had an unlikely ally when a Government Minister stood up and gave the following speech: 'Ultimately, the air we breathe, the water we drink, the food we eat and the energy which powers enterprise, are all threatened if we do not practise proper stewardship of the planet.' The words came from Michael Gove, the new Secretary of State for the Department of Environment, Food and Rural Affairs (Defra), recently returned to the political fold following his role in the Brexit referendum. A disastrous general election campaign by Theresa May had left her with a minority government and a dwindling group of political friends, meaning Gove's right-wing appeal was needed. He was thrown a political lifeline: Environment Minister. Environmentalists including myself were concerned. His first speech, held at WWF's London offices, therefore came as something of a pleasant surprise. He continued: 'If we consider the fate of past societies and civilisations, it has, again and again, been environmental factors that have brought about collapse or crisis.' He even added, as if directly addressing his critics: 'It is because environmental degradation is such a threat to future prosperity and security that I deeply regret President Trump's approach towards the Paris Agreement on Climate Change.'

Two weeks later, Gove announced that the sale of all new petrol and diesel cars would be banned by 2040. The obvious criticism from some clean air campaigners was 'why so long?', but the importance of the announcement shouldn't be underestimated. For many environmental policy watchers, it wasn't new news – to meet the UK's 2040 carbon reduction targets, transport would be unlikely to run on fossil fuels by then anyway. But it was new news to most of the population, and the word 'ban' hadn't been uttered before. Electric cars were suddenly no longer a question of 'will it happen', but a question of 'when'. The stories about air quality and diesel that had slowly been creeping up the news agenda were now underlined with mainstream political targets. Following that announcement, in January 2018 diesel car sales in the UK dropped by 25 per cent compared to January 2017, while growth in 'alternatively fuelled vehicles' rose by 23.9 per cent.

It took a political ego like Gove's to bring clean air to the front and centre of British politics. But Defra, and the UK government, had in fact been led there kicking and screaming via a number of embarrassing court cases. The UK had been allowing its major cities to violate EU legal air pollution limits for NO_2 for years, until a then little-known environmental group ClientEarth, led by a charismatic New York lawyer-cum-Buddhist priest, James Thornton, took it to court.

I first interviewed James Thornton for the *Financial Times* in 2016. He began his career at the US Natural Resources Defense Council (NRDC) in the early 1980s, suing the Reagan administration when it briefly stopped enforcing the Clean Water Act. He set up the NRDC office in LA, where he also trained in Zen Buddhism and became pally with some green Hollywood A-listers including 'Leo' (Leonardo DiCaprio). He initially moved to the UK to take a break from legal activism, teach meditation ('the Dalai Lama asked me to do that'), and marry his English husband – then not an option in the US. But his legal itch soon needed scratching. He retrained in European law and set up ClientEarth as an NRDC equivalent in the UK (a Hollywood friend, Emma Thompson, lent him a room in her West Hampstead flat as his first office). 'There are about 15,000 corporate lobbyists in Brussels, many with very highly paid lawyers,' he tells me in a soothing yet direct voice, which I imagine works equally well for both Buddhist and courtroom duties. 'There was a great inequality of arms. Our idea was to balance that ... When I first came up with this idea of bringing an air quality case, there was nothing in the papers, even major environmental organisations weren't interested.' The EU's Air Quality Directive (1996) gave member states until 2010 to reduce the concentration of NO_2, with legal eight-hour and monthly limits. The 2010 deadline was fast approaching, yet there was very little action to be seen. The UK government responded to ClientEarth's first information request by admitting it had no intention of meeting its target. So, Thornton sued.

His first case in 2011 saw the High Court rule that the UK was indeed in breach of the EU directive. But it didn't specify what the government needed to do about it, so again the government chose to do nothing. The second case in 2013 at the Court of Appeal also sided with ClientEarth, but this time passed the buck to the European Court of Justice (ECJ). 'In EU law, human health takes paramount importance – it takes precedence over all other considerations including economic costs,' says Alan Andrews, Thornton's right-hand man on the air quality case, picking up the story. 'The ECJ gave a landmark ruling which is now the leading precedent for air quality law across Europe. We see the right to clean air now enshrined within EU law … It is an obligation on member states. It is no longer a case of trying to comply – you *must* comply.' The 'final' win in October 2016 ruled that the 'Secretary of State had fallen into error', and must speed up all plans to comply with the EU law. Furthermore the 2015 plan was described as 'unlawful'. That day Prime Minister Theresa May announced in the House of Commons that the government would not appeal, and would comply in the shortest time possible. Not a bad result for a rabble-rousing New York Buddhist. According to Thornton in mid-2017, Defra under Michael Gove is no longer 'disputing that you need to clean up the air; the government before did dispute that'.*

The UK's first ever National Clean Air Day (NCAD) was held on 15 June 2017. It wasn't exactly celebrated with royal anniversary-style street parties, but as I took the train to London, early signs on Twitter were promising. A collective effort including my own tweets had got #CleanAirDay trending UK-wide, and I started to see pictures from Leeds of parked cars with flowers emerging out of their engines, and from Manchester of the mayor, Andy Burnham, visiting a pop-up 'lung information dome' in the city square.

Arriving in London, after a train and a tube, I emerge out of Archway station a little disorientated. I lived in this area for years. The sight of cars and lorries trudging wearily up the A1 was a virtual 'welcome to Archway' sign. The giant roundabout at the top – so big that it was technically a

---

* Even so, just to make sure, ClientEarth returned to court in February 2018, securing a ruling that the High Court should have 'effective oversight' of the UK government's next air pollution plans. The judge said: 'The history of this litigation shows that good faith, hard work and sincere promises are not enough … and it seems the court must keep the pressure on to ensure compliance is actually achieved.'

'gyratory', apparently – was infamous among the capital's cyclists. In 2013 there was more than one serious accident every month involving cyclists. But the gyratory has – I now realised – been replaced by a huge, permanent pedestrianised zone with cycle lanes. The roundabout island, previously home to the apocalyptically unenticing 'Dusk Till Dawn' bar, has now rejoined dry land (the bar reborn as the only slightly more appealing Tropical Bar). The cycle lanes hosts a trickle of grateful cyclists, like war survivors returning home from the trenches – albeit only for a short while, before battle recommences up the cycle-lane-free Highgate Hill.

A short way up the hill at Whittington Hospital, I have arranged to volunteer with the Islington Council 'no idling' team: walking up to people in parked cars and vans with engines still running (i.e. 'idling') and asking them to turn their engines off. The group of volunteers gather ready to go out onto the roads, and I'm handed a blue high-vis vest with 'Switch off engines for clean air' written on it, and paired with Jo, from the council's clean air team. I'm wholly expecting open hostility from the drivers, but I'm actually surprised by how polite all the exchanges are. We ask if they would mind turning their engine off, and would they like a leaflet, and largely hear 'Ah yes, sorry, I'll turn it off, no problem.' Would you like this educational game for your children? 'Yeah, OK, why not. Cheers.' Most people don't actively, aggressively pollute. We simply forget how pollution is caused.

After my 'no idling' shift I meet with Victoria Howse, who boasts the wonderful job title 'ZEN Manager' – ZEN, in this instance, not being the James Thornton kind, but 'Zero Emissions Network', an air quality collaboration between the London boroughs of Islington, Hackney and Tower Hamlets. She is manning a stand on the other side of Archway tube station, offering free bike repair and free coffee (the queue is longer for the coffee). The ZEN, part-funded by the Mayor's Air Quality Fund, is 'trying to work with businesses to get them to switch to sustainable travel methods with lower emissions', Victoria tells me. 'We're hoping that the removal of the gyratory and the new cycle lanes will be a big improvement.' When I leave her and go on to Shoreditch I find a 'pop-up' wooden-clad green space of benches, trees and plants, plus bicycle racks, inviting walkers and cyclists to sit and rest. It's actually on the road, taking up two parking spaces, literally reclaiming the tarmac from cars and returning it to the public realm. Laura Parry, the second of two ZEN Officers, meets me at the 'parklet'. 'This area will soon be an "electric street", where only zero-emissions-capable cars can go down … It's currently where two of the worst air quality readings in the Borough are.' I ask whether the café opposite minds losing its parking spaces. 'For them it's really good because it basically creates extra outdoor space for their customers. Around here there are multiple places for parking, so it is not affecting things like deliveries. But also with the changes like the ULEZ [Ultra Low Emissions Zone] and the Electric Streets, it's not going to be practical for local businesses to be using polluting vehicles anyway. So, we are helping businesses swap over to cargo bikes, or electric bikes, or EV [electric vehicles].'

I end National Clean Air Day in Greenwich, south-east London – the opposite end of town from where I had started. The Greenwich Centre has a shared district heating scheme – private apartments, a library, leisure centre and swimming pool are all heated by a single large gas-powered boiler. Its CO_2 and NO_2 emissions are much lower because of this. At 5.30 p.m. Dan Thorpe arrives: a local primary school teacher by day, and elected Borough Councillor the rest of the time; as he strolls into the Greenwich Centre and glad-hands a few people he recognises, I realise who has had the longer day.

Greenwich has some obvious air pollution hotspots, most notably the city's busiest traffic tunnel under the Thames, the Blackwall Tunnel, and London's only major car ferry, which crosses the Thames at Woolwich. Dan helped create the local Air Quality Action Plan, and successfully bid for funding to implement a Low Emission Neighbourhood area that will bring £2 million of investment for schemes such as car-free days and electric vehicle grants. 'When I became a councillor in 2004, environmental issues were not really on my agenda,' he admits. 'In 2014 when I joined the Cabinet, there was growing awareness around the issue of air quality … the deaths attributable in our borough are somewhere in the region of 600 a year. Any death is one too many. That has focused people's minds.' In Mayor Khan's new London, the fightback has begun.

Beijing's Olympian feat

In the spring of 2014, China's President Xi Jinping stepped outside for a photoshoot in the smog. Chinese anthologist Jerry Zee later described its significance as a moment when, 'the highest official in China allowed himself to be captured throwing in his lot with the common Beijinger … held together not by citizenship but by shared exposure to toxic weather.' Beijing residents had flooded social media with facemask selfies, holding placards reading '#I don't want to be a human vacuum cleaner' (#我不要做人肉吸尘器), and were getting a response. Xi Jinping's number two, Li Keqiang, gave a speech later describing air pollution as 'nature's red-light warning against inefficient and blind development'. Live on Chinese state television he declared a 'war against pollution'. It was a game changer. Li immediately outlined the measures to be taken: reducing PM and eliminating outdated power stations and industrial plants. China cut steel production capacity by over 27 million tonnes (the equivalent of Italy's entire output), slashed cement production by 42 million tonnes and shut down 50,000 small coal-fired furnaces. Li promised to change 'the way energy is consumed and produced' and to promote green and low-carbon technology. Beijing, until recently reliant on coal for its energy, closed its last remaining coal-fired power station in 2017.

The Beijing-based Institute of Public & Environmental Affairs (IPE) is a short but confusing walk from Jianguomen Metro station, and when I arrive at the tower block there is no company or organisation branding, simply unmarked doors and an elevator. I phone to check I am at the right place, and am told to come up a few floors, where I'm met by a young American who speaks fluent Chinese, Kate Logan, a director at IPE and a board member of the Beijing Energy Network. She offers me the customary drink of steaming hot water. IPE was founded in 2006 by Chinese environmental activist Ma Jun to promote information transparency and public participation in environmental reporting, and was China's first genuinely independent environmental non-profit. IPE began by collecting public (but often obscure) environmental information and collating it into an easily accessible database. This became the China Water Pollution Map, and later a broader database of environmental violations including air quality information known as the 'Blue Sky Roadmap'. I'm intrigued to know how IPE got away with openly criticising the government in the days when air pollution was not officially recognised, publishing statements such as 'large cities are increasingly suffering from the rapidly expanding and serious problem of air pollution' and criticising 'China's lax environmental supervision and the low cost associated with violating' as early as 2011.

'In the beginning it was very sensitive in a lot of ways,' admits Logan. 'I think when IPE first started producing the database, the specific decision was made to only take information from government sources or officially verified sources. That was basically a means of mitigating that sensitivity ... if anyone came doubting the validity of the data, you can verify the source because it's an official source. That led to a certain level of credibility ... if you look at the trajectory of environmental laws and policies in China you will see a lot of trigger events significantly influencing a government policy response.'

There were three main trigger events, says Logan: the 2008 Olympic Games; the US embassy air quality data; and the 'Airpocalypse' smog of 2013. Each of the three built upon the others to force Xi Jinping out onto the balcony that day in 2014, and forced air quality up to number one in the political agenda.

The run-up to 2008 Beijing Olympic Games was plagued with international concerns about the smog that awaited the world's visiting athletes. Beijing was already synonymous with smog, and with one year to go, the Olympic president Jacques Rogge was forced to reassure invited nations that certain events could be moved to new venues or dates on the smoggiest days. Reuters described it as the most 'intensely scrutinized preparations for any Games in Olympic history', and China was going to do everything within its power not to be embarrassed on the world stage. The actions it took inadvertently created a blueprint for the city's future. With a month to go until the opening ceremony (on the auspicious date 8/8/08) the Beijing Dongfang Chemicals Factory, Yanshan Petrochemical Company, Shougang HONGYE Steel Plant and Beijing Flat Glass Group Corporation either halted or limited their production. Four major coal-fired power stations began to use only low-sulphur coal and operated at 30 per cent capacity. All cement companies temporarily halted production. Industrial production was also slowed or halted in a vast area surrounding Beijing, including Tianjin, Hebei, Shanxi, Shandong, and as far as Inner Mongolia some 200km (125 miles) away. Open field burning was banned across the whole of northern China from May to September. From 1 July to 20 September, all vehicles that did not meet European Euro 4 emission standards were banned from entering the city, and only half of those that did meet the standards were allowed in each day via an odd/even number-plate system.[*] In the first 20 days of August, sulphur dioxide, PM10, carbon monoxide and nitrogen dioxide decreased by 27 per cent, 40 per cent, 50 per cent and

---

[*] The first major city to use the odd/even system was Mexico City in 1989. When in operation, only cars with a number-plate ending in an even number can be used within the city on one day, followed by cars ending in an odd number the next day. As a short-term measure, it works really well. As a long-term measure, not so well, as Mexico City found out when the middle classes simply began buying two cars, meaning the overall car density – number of cars per head of population – went up.

61 per cent respectively, compared to the same month the year before. August 2008 was Beijing's cleanest month for 10 years (and back then, in 1998, Beijing had roughly 7 million fewer inhabitants).

The Olympics experience proved to the authorities what could be done with the right will and regulation. It also proved beyond any reasonable doubt where the emissions had come from in the first place.

Then the US embassy in Beijing began publishing its own PM2.5 data. 'That was a turning point that definitely influenced the speed at which policies were drawn up,' says Logan. I hear this more than once during my visit to Beijing. Mary (her Anglicised name), a highly educated business woman from Tianjin and former university lecturer, told me that no one used to know what the air pollution was, including her. 'I just thought it was fog. But it had a different colour, and different taste. Then the US embassy released the figures about something called PM2.5. And suddenly we had a name for it. And everybody started searching on the internet about "PM2.5" and learning what it was. That was a big change,' she says. 'Now everyone knows what PM2.5 is.'

On 1 January 2012, Beijing published its very first official trial PM2.5 data on the China National Environmental Monitoring Centre's website, from a single monitoring site at the centre's observation laboratory. By doing so, it became the first city in China to publish official PM2.5 data. Suddenly the monitoring sites started popping up all over China. On 8 March, Guangdong province published information from 62 monitoring sites in nine cities. By October, Beijing had already upped its official monitoring stations in the city to 35. Watching the expansion of transparent air quality data in China, says Kate, was like seeing one little red light appear on the map, 'and one at a time lots of little lights start to pop up … first it was Beijing, then four cities, and then it jumped up to 170, 380 … now it's across China.'

In 2014 the government made real-time industrial emissions data from 'around 13,000 factories across China required to be publicly disclosed for the first time', Logan informs me. 'I think when the government started to realise that what IPE was doing was essentially helping to support the government's supervision efforts ... that was a shift in mindset. And because the pollution issue got so bad that the government basically knew they had to start doing something, that showed they were actively responding.'

With Beijing's pollution now widely and accurately monitored, there was no hiding place from its worst pollution episode in history – the 2013 Airpocalypse (described in Chapter 1). Then came *Under the Dome*, an independent documentary film by former TV journalist Chai Jing, to underline the sheer levels of pollution the country was experiencing. Released on 3 March 2015, it was reportedly viewed 300 million times before it was taken down by the authorities four days later, on 7 March. In a very similar format to Al Gore's *An Inconvenient Truth*, Chai Jing's film took the form of a mixed media lecture, interspersed with short films of her investigations across China. It pulled no punches. Pollution was shown to be an obvious result of China's dependence on coal, cars and steel, and its inability to enforce what had become lax regulations. 'A lot of people knew air pollution was there,' says Logan, 'they knew it was bad, there was beginning to be access to data ... but how to protect yourself or how bad it really is or what the contributing factors are – I think *Under the Dome* answered those questions for the first time ... if you look at the reasons why China has made such a big push to have better environmental governance it is basically because it threatens public health and then also threatens social stability.'

Rather than attempting to suppress or hide pollution data, now China is arguably going the other way. The Ministry of Ecology and Environment now has a five-digit hotline for members of the public to report any environmental transgression, via text message and WeChat. IPE's own Blue Sky Roadmap app is now also used by the authorities as a public reporting mechanism, and departments are required to respond within seven days to any report made via the app. 'It empowers the public domain,' says Logan. 'China is way ahead of many countries in environmental reporting and especially real-time emissions data. These sort of mechanisms are starting to come about as a result of this idea that the government needs to have better channels for communication and interaction with the public ... but at the same time diffusing any sort of tension that will snowball if you had too many people for the government to be able to handle at the same time. Hence *Under the Dome* being censored.'* The red alert, the most serious warning level on the Chinese system, was first issued in December 2015. When pollution reaches this level, Beijing in effect implements the 2008 Olympics measures: polluting industries are shut down and vehicles are rationed, while pollution alerts are widely circulated on apps and social media (including Weibo and WeChat), television and radio.

James Thornton is now working with the Chinese judiciary to help strengthen environmental control laws, including the training of judges. 'China is ahead of the UK in many respects,' he tells me. 'They have woken up to the fact that they need to clean up, and they are taking clear and dramatic action ... they passed a law in 2014 that allows citizens to bring cases in Chinese courts against polluting companies. That is a deeply democratic operation ... The government is incredibly eager to make the system work. They have closed down over 1,000 coal-fired power stations across China ... and have actually changed the Chinese Constitution to articulate that they are building what they call "an ecological civilisation".'

---

* I heard from more than one source, albeit impossible to verify, that *Under the Dome* was released with the full knowledge and permission of the Chinese censors. The theory was that it caused just enough outrage to help push through greater emissions control and enforcement powers, but not enough to spill over into broader unrest.

While in Beijing I am invited for lunch with a former Chinese ambassador to several major Western countries. I can't use his real name, though I verified his identity prior to and following the interview. His Communist Party seniority makes me nervous, and I find myself fumbling with my chopsticks, but he is of course – being an ambassador – both polite and charming. The ambassador tells me that Beijing used to suffer from terrible sandstorms. The trees in the surrounding area had steadily been chopped down for firewood and building material. Trees of course provide a natural windbreak (and hold the soil together, preventing landslides) but after decades of clearance, Hebei province had become a dust bowl. The dust would blow into Beijing, choking its residents. A grand programme of replanting the green belt then began. China even has a national tree planting day on 12 March. President Xi Jinping is eagerly photographed with a spade and a sapling every year. In March 2017 he told Chinese state media, 'We expect blue sky, white clouds, clean water and fresh air, which are all related to ecological construction. The people should live in green shade, and this is the target of our efforts.' The restoration of the green belt worked. Beijing no longer suffers from dust storms. In 2018, China went a step further, assigning over 60,000 soldiers to plant at least 84,000 square kilometres of trees (32,400 square miles, roughly the size of Ireland) by the end of the year, the majority in Hebei province around Beijing.

Another blight Beijing has removed is coal. The Ambassador reminisces how coal used to be the only fuel people used. Back in the 1980s, he remembers that briquettes of coal – made using the lowest grade of reformed coal dust – used to be a common sight on the back of bicycles. People would use them for the dual purpose of cooking and heating. Every home would be the source of coal smoke. By the 1990s, coal-fired powered stations were then added to the mix, many situated within the city itself. But coal too has since been dealt with, replaced in both homes and power stations by natural gas. The Ambassador tells me that coal mines are now often drilled only for their pockets of natural gas; the coal itself is increasingly left in the ground.

A Chinese tech entrepreneur, who also asked not to be named, told me, 'One of the things that only China could do is they built a bunch of new [coal-fired] power plants in 2014. In 2015, they decided to get rid of them. They spent hundreds of millions of dollars, they had been in operation for three months, and then said "OK, new law, no coal power plants". So now they are just being knocked down. That could not happen anywhere else in the world. And you could say it is a terrible decision, but it is also a great decision because it is the only way to fix the problem … They just decided they were going to tackle it. It is no longer a sensitive subject, it is something that we can change and be proud of, and show the world. Just like that. No more coal in Beijing.' That sounds like an exaggeration, but the two plants he refers to were the tip of an iceberg. In October 2017 it was reported that China was to stop or delay work on 151 planned and under-construction coal plants, with capacity totalling 95,000 megawatts (equal to the combined operating capacity of Germany and Japan). In the year to that point, natural gas consumption had already increased by 17 per cent, replacing an estimated 47 million metric tonnes of coal.[*] The province of Shanxi shut 27 coal mines in 2017 alone, while Taiyuan also banned the sale, transport or use of coal by individuals or small businesses, all for the sake of improving air quality.

By January 2017, Beijing's acting mayor, Cai Qi, announced a crackdown on 'open-air barbeques, garbage incineration, biomass burning, and dust from roads', according to the

---

[*] Although one thing the Chinese Ambassador did tell me 'not to get too carried away with the numbers, when talking about China'. Forty seven million tonnes is a drop in the ocean when China consumes close to four billion tonnes of coal a year. Coal still made up 62 per cent of China's total energy consumption mix in 2017.

state-run Xinhua news agency. Chinese Premier Li Keqiang promised at the annual National People's Congress that year to 'make our skies blue again'. By 2020, the city plans to replace more than 70,000 petrol and diesel taxis with electric vehicles and install 435,000 charging stations. Starting from 2018, the Chinese government requires one out of ten vehicles manufactured there to be electric. The official goal is the production of two million electric vehicles (EVs) a year by 2020, and seven million a year by 2025 – which would mean that by 2025 electric vehicles account for around a fifth of the total auto production of the country. As it stands, according to the China Association of Automobile Manufacturers, there were around 424,000 'new energy vehicles' produced in China during the first three quarters of 2017 – which represents a 40 per cent year-on-year increase compared to the first three quarters of 2016.

Few stones are being left unturned in China's war on pollution. Almost every building in Beijing (and most cities in China) is heated by district heating, from a residential tower block to the Presidential Palace. Huge boilers fire up on the same date every year, heating people's homes and offices for an allotted time every day. It is already a very efficient system, one that many Western cities are keen to copy. Beijing's boilers used to be coal-fired, but are now all gas. But even so, gas boilers produce NOx, and the Beijing authorities want to reduce NOx still further. In Beijing I accept an invitation from the US company ClearSign to visit a pilot project retrofitting district boilers to make them produce far less NOx. Each boiler is the size of a bus, and in Beijing alone there are approximately 3,500 of them. A boiler is effectively a tube with a large flame at one end, which heats up lots of water pipes that wrap around the tube to produce hot water or steam. In a typical boiler, the flame is a large flickering one that causes pollutants at high temperatures. The ClearSign retrofit installs ceramic tiles full of lots of little holes halfway up the boiler tube. The flame hits the tiles, and the tiles distribute the heat more effectively, turning one large flame into hundreds of tiny ones.

ClearSign's man in China, Manny Menendez, picks me up from my hotel in the central business district and, as his driver takes us out to a residential area, we have a long chat on the back seat. I quickly realise there aren't many people in the world like Manny Menendez. He was doing business in China before the rest of the world thought it was possible to do business in China, becoming friends with the then-reformist leader of China, Deng Xiaoping. The car pulls up at the metal gates of a grey Beijing District Heating Group building, near some large residential tower blocks. The lot is strangely quiet, and Manny walks confidently up to an unmarked door and pushes it open. The chill of early winter is just as present inside, so we keep coats and scarves on. A security guard welcomes Manny and his interpreter warmly, and we go through to the boiler room where four large tubular boilers, each the size of a minibus, stand side by side. Three are circled by large red metal flue-gas recirculation tubing, the current industry standard for NOx reduction, which recirculates some of the air from the chimney or 'flue' back to the flame again, which lowers the flame temperature and reduces oxygen content, lowering the NOx. The fourth boiler, retrofitted internally with ClearSign's ceramic tiles, stands relatively naked. Despite the fact that we are standing so close to huge gas flames, this room is no warmer – all the heat is going into the water pipes.

We leave to talk in a nearby workers' café, over a bowl of pepper soup. A kung fu movie plays loudly on a television mounted to the wall. 'That building only has four boilers,' says Manny. 'Within the Beijing District Heating Group, the ones that are in their direct ownership and control, there's 1,500. Then through joint ventures or partnerships with others … there are another 2,000 … So we just saw a 29MW boiler. But in one of the other facilities I have seen 116MW boilers, much bigger … they also have a huge inventory of smaller 14MW boilers… We've already looked at those – they have 75 of those ready to go, and we will go through those one by one.' He tells me that the Chinese NOx regulations have been getting lower year by year. 'A few years ago in the cities it was in the region of 60–70ppm, and it is now down to 30–35ppm. In District heating now they have to meet 15ppm. And the new burners they are putting in from European and US companies require additional equipment, flue-gas recirculation [FGR], in some places FGR plus selective catalytic reduction – and there are issues with efficiency, maintenance ... using things like urea and ammonia ... It costs more and more money to operate ... Whereas we don't need any of that, and are getting 5ppm or less. The project we have in the States at the Exxon-Mobil Shell, they are getting 2.8ppm of NOx – it's unheard of.' Then we get back into the car, and Manny tells me about the time he played football with Franz Beckenbauer.

Delhi denial

Delhi almost won its battle against air pollution before most cities had even woken up to theirs. In 2003, Delhi was awarded the Clean Cities International Award by the US Department of Energy in recognition of its 'recent improvements in air quality'. Between 2001 and 2002, all commercial passenger vehicles – buses, taxis and three-wheeler 'rickshaws' – were ordered by the Indian Supreme Court to convert from petrol and diesel engines to the cleaner fuel compressed natural gas (CNG). By the end of 2002, the conversion was swift and almost complete, totalling an estimated 15,000 buses, 55,000 three-wheelers and 20,000 taxis. According to the Energy and Resources Institute, diesel buses in 2002 emitted 54 times the amount of PM in grams per kilometre than a CNG buses, and 2.5 times the amount of NO_2 (although the CO emissions of CNG were twice as high). After the implementation of CNG, the particulate levels dropped in the city by about 24 per cent compared to 1996 levels. A study by Jawaharlal Nehru University found a perceptible drop in polycyclic aromatic hydrocarbon (PAH) emissions. The Delhi Metro system, which began construction in 1998, also opened its first line in 2002, and by 2006 it had three underground electric train lines spanning much of the city. Delhi could briefly lay claim to having one of the cleanest public transportation systems in the world. Having solved its air quality problem to international acclaim, how then, just ten years later, did it become the most polluted major city in the world?

The Central Road Research Institute (CRRI) in Delhi is a veritable palace of bureaucracy – a huge stand-alone white and red building that spreads for hundreds of metres. Inside I meet one of its many princes, Dr Niraj Sharma, senior principal scientist, environmental science division. A larger-than-life, welcoming character, he has a booming voice and a ready smile. During our conversation, which lasts for almost two hours, a steady stream of subordinates come in with papers for him to sign, while another is summoned to bring us coffee and potato samosas. It is not yet 10 a.m., and I have just eaten a large cooked Indian breakfast. But he insists that I go ahead, and any pause in eating – or in drinking the coffee – causes a concerned pause in conversation and 'please' as he gestures to the cup or plate.

'Most other countries in Asia suffer from a lack of democracy. We suffer from the excess of democracy,' he laughs. When the Supreme Court ordered the mass conversion to CNG in 2001 'there was a lot of public demonstration against this', he tells me. 'Police authorities were given full authority to control the mob if they came within 100 metres of the Supreme Court.' By his estimation, 'around 1 Lakh [100,000] auto-rickshaws were converted into CNG ... From 2002 to 2006–7, things were quite good. Then we started introducing Euro 1, 2, 3, 4, everything was introduced. Then what happened, probably you will be surprised to know – how much do you think the registered vehicle population is in Delhi? Any idea? Approximately 10 million ... more than the combined vehicle population of Mumbai, Calcutta, Chennai and Hyderabad. All of these four cities vehicles combined is less than the number for Delhi. Whereas the human population ... Is it too spicy?' He stops, looking concerned as I struggle to finish the first giant potato samosa. I shake my head and do my best to munch appreciatively. 'In

Delhi,' he continues, reassured by my chewing, 'the vehicles are growing at the rate of approximately 10 per cent per annum. Every day around 350 vehicles are added to the fleet. We have reduced the pollution coming out of each vehicle significantly. But that has been compensated by the increasing number of vehicles, and increasing the travel distances of each vehicle. So, whatever we gained, through CNG, through vehicle technology, through improved fuel quality ... that has been mitigated by the increasing number of vehicles, and increasing the travel distances of each vehicle.' When I can't quite face the second potato samosa, I fear I have offended him. I arrive at my next meeting with the second samosa – skilfully wrapped in a sheet of A4 printer paper by one of Dr Sharma's courtiers – in my laptop bag, for my lunch.

Delhi, like London and Beijing before it, is on the cusp of waking up – again – to its air quality crisis. But many parts of its society are happy to keep slumbering. Jyoti Pande Lavakare is an economist and air quality activist. Having worked for the Dow Jones South Asia bureau and lived in America for a time, she was shocked by how much worse the air pollution had become when she returned home in the mid-2010s: 'My friends in Delhi were saying, you've become a total American, get real, you're fine, we're fine. I started to question myself, am I over-reacting, is the air quality not so terrible? So, I decided to research this a little more to find out for myself ... And I was appalled to find out it was worse than I thought. I started to talk to other mothers about this.' Together they formed the campaign group Care For Air, and started visiting schools to give presentations and raise awareness. 'We would tell them about air pollution, have a question and answer session ... We found a lot of myths – that somehow being Indian you are above it all, that "my lungs are strong, I am Indian, I can deal with it". Which is a complete myth, and now more and more research shows that actually Indians have lower lung capacity than average due to living in more polluted conditions ... I feel we are still at a very early stage of awareness. It is like two steps forward, one step back.'

The steps forward include the 11 November 2017 front page of the *Delhi Times*, featuring 12 of the capital's most highly respected senior doctors. In a striking image they pose in front of a smog-shrouded parliament, each wearing whites and a stethoscope, under the headline quotation: 'NO ONE SHOULD BE LIVING IN THIS CITY'. Part of the caption beneath reads, 'the doctors are unanimous: It's an emergency'. Another member of the mothers' group, Shubhani Talwar, tells me how it happened: 'We called them and called them, and the 12 doctors said our passion was so infectious that they said let's get on with it. They couldn't say no.' She didn't know exactly who would turn up until the day of the photoshoot. 'Then one came, then another came … the best doctors of Delhi, they all stood with us. We didn't want to make it political. It is all about medical issues. It doesn't matter which political party you support all of you sort it out, because mothers and doctors are just letting you know that what you are doing is not feasible.'

My B&B host during my stay, Vandana, tells me that 2017 felt like the first year there had been a real political argument about air pollution (although she admits she stopped watching the news about four years ago because 'it got too depressing'). The sale of firecrackers at Diwali was a turning point, she said. Two to three years ago, there was a school movement to stop burning celebratory firecrackers due to the smoke pollution. Diwali celebrations reach a climax on the day itself, but for two or three weeks there are fireworks, and firecrackers make for unbearable levels of pollution. The street dogs began dying, said Vandana, who looks after some of the dogs herself. Visibility could go down to near-zero. This year, however, she saw far fewer firecrackers. The government has finally banned the sale of them.

But then there are the steps back. Everyone agrees that the autumn stubble burning of crops is a major contributor to the smog from September to December. Yet criticism of farmers does not play well politically. In October 2017, chief minister Captain Amarinder Singh announced that the Punjab government would not penalise farmers for breaking the ban on stubble burning. Of the 6,670 cases of stubble burning recorded that season, 80 per cent were reported to have occurred after his statement. An unnamed environmental official complained to the *Hindustan Times*, 'Politics has once again derailed our drive against stubble burning ... farmers are setting straw on fire every night.'

Alongside the stench of stubble burning there is the whiff of denial in the Delhi air. For every minister or ministry making bold and progressive moves – such as the Road Minister declaring an ambition to become a '100 per cent electric [car] nation' by 2030 – there are others doing the opposite. A press release issued by the Indian Ministry of Earth Sciences on 1 February 2014 claimed there was 'no systematic increasing or decreasing trend in the level of PM2.5 during past 4 years in Delhi', despite abundant evidence to the contrary including, astonishingly, a report that same month – February, 2014 – from the Environment Pollution (Prevention & Control) Authority for the National Capital Region, which stated: 'The annual average PM10 levels were reduced by about 16 per cent between 2002 and 2007 [due to CNG]. But thereafter, with rapid motorisation, particulate levels ... increased dramatically by 75 per cent. Between 2002 and 2012, vehicle numbers increased by as much as 97 per cent, contributing enormously to pollution load. Moreover, between 2002 and 2011 the nitrogen oxide levels have also increased 30 per cent indicating Delhi is in the grip of a multi-pollutant crisis.' Despite this clear detail, the Indian Ministry of Earth Sciences blamed high pollution episodes on 'untimely synoptic weather' and sudden changes in wind direction. The fact that such untimely changes seemed to happen every Diwali and stubble-burning season, or even every rush hour, didn't warrant a mention.

The disproven claim that it is 'just the weather', however, remains popular. I heard it so much that I started to call it 'Delhi denial'. When I interview Professor Mukesh Khare at the Indian Institute of Technology (IIT), despite all the facts and figures he gives me about pollution from diesel lorries and stubble burning, and his belief that 'penalties on second cars should be very heavy', he *still* says: 'It is a meteorological smog, it is not a photochemical smog, it is not a source-level smog … Meteorology is making this smog happen in Delhi. There's a lower mixing height, lower temperature, calm winds. You can't control meteorology.' And Dr Niraj Sharma at the Central Road Research Institute, who has spent 25 years monitoring vehicular emissions, also can't help but say, 'I am a firm believer that it is mainly meteorology – wind speed and wind direction along with rainfall – which controls the air pollution … In my opinion, dust is bound to happen in India. Because the wind is blowing from the Pakistan and Arab side, it contains a lot of fine particles. Once the small particles are there they remain in the atmosphere … It enters into Delhi. And my perception is dust in India is a natural phenomenon.'

The data clearly show how Delhi's daytime peaks of PM2.5 (which includes transboundary 'natural dust') coincide with the morning and evening traffic rush hours. Cars are not blowing in from the Pakistan and Arab side. The coal smoke from Delhi's inner-city power stations cannot be blamed on near-neighbours either: in 2014, 75 per cent of India's electricity production came from coal, compared to hydroelectric-rich Pakistan's 0.15 per cent. A major 289-page study of air pollution sources in Delhi by IIT, between November 2013 and June 2014, found that on average vehicle exhaust emissions account for 25 per cent of PM2.5 and above 35 per cent in certain locations, the remainder coming from road dust (38 per cent), domestic fuel burning (12 per cent) and local industry, including power generation (11 per cent). NOx emissions are even more localised, with 52 per cent of emissions from the local industry and power generation, and 36 per cent from vehicular emissions that, says the IIT report, 'occur at ground level, probably making it the most important emission'. The city's sulphur dioxide emissions come almost entirely from Delhi's coal power stations and the roughly 9,000 hotels and restaurants that burn coal in their tandoor ovens – the traditional clay oven that uses charcoal or coal, similar to a Western barbecue. As for 'agricultural soil dust', the most commonly blamed external problem that supposedly blights Delhi, the report concluded, its contribution to PM2.5 'is negligible'.

I decided to visit the Mexican embassy while in India, too, because its ambassador Melba Pria has been an outspoken advocate for clean air in the city since her appointment in 2015. She immediately caused a stir by refusing an ambassadorial car, preferring instead to drive a three-wheeler CNG rickshaw. By highlighting the issue of vehicle pollution, she became known as much for her clean air activism as for her ambassadorial duties. But she has always been keen to stress that her interest is not that of a wealthy Westerner hectoring a developing nation – but that of a similarly large, developing country, that had previously had an extreme air pollution problem, and has emerged with some solutions to share. The embassy sits within a wealthy gated community of largely residential houses, and my Ola driver (the Indian rival to Uber) is not allowed in, so I take the opportunity to walk a few blocks. Unlike most of the city, its streets are relatively car-free and easy to walk down. The only people out on the street are the servants and labourers, who build and clean and guard these neighbourhoods, before returning to their homes – often at the very edges of the city – at night. The houses of the rich hide behind tall electric gates, while those under construction give off a mist of toxic dust from uncovered piles of cement that stand by the road.

Ambassador Pria is a striking presence, with floating long linen robes and free-flowing long hair with grey streaks. 'I have been an activist on good air because I lived through that,' she says of growing up in Mexico City. 'We are just another country that has had various periods of pollution – we were there in 1992 … we were the worst city in the world, as maybe Delhi is today … I've been very vocal, *very* vocal about air pollution. Because of my using of the auto-rickshaw, I am very popular. But do you think that one person of the Delhi government has called me? Many of Mexico's solutions will work here … But not one person from the local government has called me.'

I put some of the 'Delhi denial' comments I've heard to Ambassador Pria, and she shakes her head in rueful recognition. 'I shouldn't say this because I am not here to criticise anybody … But the other day someone said to me "it is karma". No, no – pollution isn't karma, we made it. It is the bad policies, bad practices … In Delhi we consume per capita the biggest amount of electricity in India … 80 per cent of that electricity is given to us by coal. In the middle of the city you will see the coal power stations.' Of the two coal-fired power plants within Delhi's city limits – Indraprastha and Badarpur – the latter is rated by the Centre for Science and Environment (CSE) as the most polluting in the whole of India. Then there is Diwali, when 'every official air pollution monitor in Delhi registered 999 [$\mu g/m^3$]', says Ambassador Pria. 'All 60 of them. Two years in a row. We don't know if it was 1,242 [$\mu g/m^3$] or 1,395 [$\mu g/m^3$] – we don't know, because it reached the [three-digit] maximum of 999 … If it was just the weather, why does it happen the day after Diwali, and not the day before? Or a week before? This year yes, it is true we had sandstorms from the east, and cold air that depressed it down … But there is no consciousness that we have to change our lifestyle.'

Another contributor to the smog is diesel lorries, many burning the worst, low-grade agricultural diesel. The major highways of northern India run through Delhi, meaning many trucks are passing through on the way to other places. Such lorries are banned from driving through the city during the daylight hours of 5 a.m. to 10 p.m. The effect of this, however, is the nightly procession of lines of lorries through the city. Professor Khare, who has studied the impact of these trucks, tells me that PM2.5 levels can therefore be even worse at night. 'They contribute a lot to Delhi's pollution at night-time,' he says, adding that there is 'a policy gap' that 'is not benefiting the Delhi atmosphere' (which again is not a meteorological problem).

Rana Dasgupta, the British-Indian biographer of Delhi, has lived there for the past 17 years, but when I visit his home he is in the process of packing to move to the US. Dasgupta's 2014 book *Capital: A Portrait of Twenty-first Century Delhi* paints an unsparing portrait of a city in crisis. His book didn't cover air pollution, but he admits that were he writing it today it would be a major focus. 'My suspicion is that it is an extremely elite concern,' he says. 'They are outraged that no one is controlling this thing on their behalf. But I suspect for a poorer majority in the city, the experience of being subjected to huge forces that they cannot control is their experience anyway.' Even among elites, air pollution is an easier conversation to have if you can blame other people or, ideally, other countries. 'In the boom years, say the late 1990s to around 2006–7,' says Dasgupta, 'young people started working in corporate jobs and earning five or ten times what their fathers had earned at the end of their careers, they had all these fancy cars, bought their parents cars, and encouraged their parents to think of consumption as a new thing … People who criticised this were treated with immense hatred … they didn't want to be told that this engine of wealth creation was in any way morally suspect or harmful to anyone. And I think to some extent this is still the case. You have people living in totally privatised universes, you have these sealed vehicles and sealed homes … So, if the air is getting worse because of this, it is not a very welcome conversation.'

In March 2017 the environment minister for the state, Anil Madhav Dave, told the Indian Parliament that 'there are no conclusive data available in the country to establish direct correlation between diseases and air pollution' and that any health impacts that were to be found were mostly caused by 'the individual's food habits, occupational habits, socio-economic status, medical history, immunity, heredity, etc.' But such denial of the facts is far from unique to India. In a similar way to dealing with addiction, admitting you have a problem is an important step to tackling it. London, Los Angeles, Paris, Mexico City and Beijing all denied their air pollution problem before finally facing up to it, so Delhi is in good company. But it's a very dangerous stage if it goes on for too long. Delhi's air pollution is arguably on its way to reaching levels of pollution unknown to a major city in human history. In the words of the *Lancet* Commission, 'rapidly growing cities in industrialising countries are severely affected by pollution' because they 'concentrate people, energy consumption, construction activity, industry, and traffic on a historically unprecedented scale'. Perhaps nowhere is this truer than in Delhi where, despite its early success in promoting CNG, its enthusiasm for the private car above all else proved unstoppable.

'People are just leaving,' says Dasgupta. 'Lots of people I know who have options in other cities or countries, especially people with young children, have decided there is nothing to do here except get out.' This now includes him. 'You can insulate yourself from lots of other things, but you can't insulate yourself from this.' To underline the point, he asks his daughter, who is playing nearby, how the pollution affects her school days. 'We have red days, yellow days and green days,' she tells me. 'On red days we're not allowed to go outside. On yellow days we can go out – like today was a yellow day – but we're not allowed to run around. And green days we can run.' How common are the red days, I ask? 'In winter they happen a lot.' I later learn that it is up to each school whether they issue any such warnings or limitations, and most do not.

Whether it is a case of one step forward, two steps back, or vice versa, remains to be seen. India is committed to cleaning up its energy supply. The Modi government has committed to install 175GW of renewable energy (100GW of this from solar sources) by 2022, more than the total electricity generating capacity of Brazil, achieving 40 per cent cumulative electric power through renewable energy sources by 2030. The days of Delhi's inner-city coal power stations, like London's before them, are surely numbered. India's Energy Minister surprised the world in August 2017 by announcing plans to make all cars electric by 2030, saying from that year, 'not a single petrol or diesel car should be sold in the country' – a full 10 years earlier than the UK. Minister for Transport Nitin Gadkari bullishly told a car industry conference, 'I am going to do this, whether you like it or not.'

At BLK Super Speciality Hospital, Dr Parakh believes: 'Public awareness is there, but only during this season, November. At this time, every news channel will be talking about air pollution ... As soon as this becomes slightly better, the media will forget about it.' But, I suggest, the true awakening needed is that even the 'clean air' season in Delhi is still very bad by international standards? 'Yes, but that doesn't happen – you're right. When it goes into the red zone, everyone asks what is happening, but nobody is doing anything. When it is still high but not alarming, then nobody is interested.'

The Californian waiver

A couple of decades after Los Angeles's first smog encounter during World War Two, by the 1960s it was a regular sight. In 1967 the California Air Resources Board (CARB) was set up – a state agency specifically tasked with monitoring air quality and regulating vehicle emissions – to fight back against this airborne menace. Three years later, the national Clean Air Act (1970) established the federal Environmental Protection Agency (EPA) to do much the same across the country. But what, then, should be done with CARB, which was already up and running? While all other states were effectively barred from setting their own emissions standards and would now have to dance to the EPA's tune, the decision was made to give California a 'waiver'. In effect, CARB could keep doing its own thing, as long as its emissions regulations were as good as – or stricter than – the EPA's national standards. This had two major effects: first, because the geography of California and LA meant it had the worst air quality problems to deal with, California's regulations have consistently been far stricter than the EPA's; and second, because California is the biggest US state in terms of GDP, the rest of the US – and indeed the world* – have in fact danced to California's tune. The 1977 Amendments to the Clean Air Act further established a 'piggyback' provision

---

* If California were a country, it would be the world's sixth largest economy, one place ahead of France, and will probably have overtaken the UK to take fifth spot by the time this book comes out post-Brexit.

allowing other states to adopt California standards instead of the EPA's if they wanted, which many chose to do.

A *New York Times* article in 1992 headlined 'California's Pied Piper of Clean Air' described CARB as 'the nation's most influential regulatory body', ahead of the EPA. 'Anything in this country with a tailpipe, smokestack or vent is likely to be regulated eventually with rules first worked out by the California Air Resources Board.' We have Californian regulations to thank for the catalytic converter,* low-sulphur petrol, PM2.5 standards, uncovering the VW scandal and the emergence of electric vehicles. Almost every step it has taken has involved tough vehicle emissions standards, and even tougher time-bound targets. The most daring and wide-reaching target was the 1990 Zero Emission Vehicle (ZEV) programme, some 20 years before the rest of the world had even heard of zero-emissions vehicles. The ZEV programme set the target that by 1998, 2 per cent of the vehicles for sale in California had to be zero emissions – a technology that was not commercially available at the time – increasing to 5 per cent by 2001 and 10 per cent by 2003. As of January 2016, approximately 192,000 new ZEVs and TZEVs (transitional-ZEVs, otherwise known as plug-in hybrids) have been sold in California, more than any other market in the world at that time. Ozone levels in Los Angeles are just 40 per cent of what they were in the mid-1970s, despite having twice the number of cars.

---

* The history of catalytic converters is another example of car manufacturers being forced to provide cleaner options by regulators, rather than the other way around. First used in petroleum refineries, it took decades for car makers to consider fitting them in vehicles. When the 1970 Clean Air Act required them to do so, General Motors, Chrysler and Ford lobbied hard against it. Ford's executive vice president said in September 1970 that it 'could prevent continued production of automobiles' and 'do irreparable damage to the American economy'. They were also successful in watering down the 1970 proposed CO and NOx standards for another 11 years, until 1981.

'The regulation of air pollution in the UK and the rest of Europe is childlike, not *even* childlike, in comparison to what goes on in California,' says James Thornton, who founded the Natural Resources Defense Council in LA. The ZEV progamme, he says, kick-started the global EV (electric vehicle) industry. 'There weren't any electric vehicles before that and as a result people started thinking about how to build electric vehicles ... When I was there working in the 1990s CARB targeted things like dry-cleaning, which released a vast amount of chemicals that caused smog, ozone, and they required them to reformulate their industry ... shops that did body-spraying of cars with highly volatile compounds in those sprays were reformulated ... Industry never does it on its own, so California regulation of air pollution is the world's best model of regulating air pollution by very intelligent and thorough standard-setting.'

Anthropogenic VOCs (volatile organic compounds) became a major focus in LA, Professor Paulson at UCLA informs me. 'Ironically, we should have very low VOCs because we don't have a ton of trees, but we add the VOCs ourselves,' she says. 'The VOCs get converted to particles later as secondary organic aerosols, which are a much bigger problem here ... because we have brilliant sunshine, lots and lots of photochemistry and not a lot of wind, all these things contribute to making our secondary smog legendary. In a nutshell, what we've done here in LA is basically everything we can think of. We have reformulated a lot of our consumer products, so we have barbecue lighter fluid that is formulated to not produce a lot of VOCs. We allow very limited use of oil-based paints, paints need to be low VOC paints ... we developed vapour recovery systems for when you are fuelling your vehicle [at a petrol station], do you have that nozzle thing in the UK?' I don't think so, I say, unable to picture a nozzle thing. 'We have this nozzle thing that comes around the tube that the gasoline or diesel comes from, and this system sucks out all of the vapours in your gas tank as you are filling it with liquid, so those vapours don't go into the air.' She adds, in summary, 'Really high NOx and VOC emitters have a hard time here.'

But California is not done yet. The state's federally approved ozone target, an eight-hour standard of 80ppb ozone, is set for 2023. To get there, NOx will have to be reduced by 70 per cent compared to 2017 levels. And given that 70 per cent of NOx in the state comes from vehicles, there is only one obvious way to achieve it: all petrol and diesel vehicles will have to be replaced.

'2023 is the target for achieving the health-based standards that were set under the Clean Air Act,' confirms Mary Nichols. 'The main thing that has to happen is the introduction of much cleaner vehicles.' Does this mean, in effect, California is transitioning from an era of cleaner fuel to an era of zero fossil fuels and zero emissions? 'Yes, correct. That is a fact. We are pushing for zero, both because of smog concerns, and because of the need to cut greenhouse emissions as well – they both point you in that direction.'

In southern California, Sam Atwood at the South Coast Air Quality Management District explains, 'Everything is going to have to transition and probably one of the toughest challenges for my agency in my region is that we have very little time to do this ... we need to transition in the region of 17 million people to zero emissions, everything from vehicles to businesses to goods movement, the whole enchilada. And you say "OK, well if we could have say 50 years to do that, 30 years?" No, our first ozone target is 2023.' To get there, says Sam, 'we offer a $7,500 rebate if you purchase a battery electric vehicle and that's offered state-wide ... The biggest bang for your buck is reducing NOx emissions from diesel trucks and other diesel equipment.' Another incentive scheme aimed at replacing off-road machinery, the $69m–$141 million a year 'Carl Moyer Program', has replaced over 61,000 high-emissions vehicles including farm machinery, construction trucks and even trains. The scheme is fully funded by proceeds from the

vehicle 'smog check' which every car needs to obtain – similar
to the MOT in the UK – every two years, or prior to a
re-sale.

In 2018, Governor Jerry Brown announced plans to spend
$2.5 billion over eight years to add 250,000 electric vehicle
charging stations and 200 hydrogen refuelling stations
by 2025. As of January 2015, Californians drive 40 per cent
of all ZEVs on the road in the United States. More than
two million new passenger vehicles were registered in
California in 2016, more than France or Spain. Major car
companies can't afford to ignore the Californian market, and
once they have remodelled their vehicles to meet its regulations,
it makes economic sense for them to sell to other markets too.
By 2017, the trajectory of EV sales was finally shooting
upwards; battery electric car sales were up 30.4 per cent
compared to the previous year – the Chevy Bolt alone shifted
13,487 units.

I ask Mary Nichols if California could also lead the way by
shifting regulatory focus beyond PM2.5 towards nanoparticles and particulate number, and she doesn't disappoint
there either. 'During my time in this field, we have moved
our focus from total mass of particulate matter, to those
under 2.5 – and … we were already looking at the ultrafine
particles when we changed to PM2.5 standards back in 1997 …
Since that time there have been more studies that seem to
confirm that the ultrafines are the worst actors.' So, do we
need new regulations and limits standards for ultrafines? 'We
might not abandon PM2.5 in the same way as we didn't
abandon total mass. We might add a new screen at the PM1.0
or under 1.0 level. Or … there may be a way to go directly to
emissions control strategies that focus on the ultra-fines,
without having to go through the years-long effort of
developing a new ambient air quality standard to monitor it …
we are increasingly looking at air pollution from a community
level, looking at total exposure of individuals … The
difference is that instead of having a network of big monitoring
stations and pursuing individual pollutants as we do today,
shifting our approach to individual exposure may change the

way we focus your regulatory activity, so we go much more directly to the source.'

Key to the Californian 'individual exposure' approach is the concept of 'environmental justice'. For decades, national and international studies have shown that poorer communities are disproportionately affected by air pollution. The link is straightforward – homes built next to busy highways, airports and industrial smoke stacks tend to be cheaper than those next to trees and parks. In recent years, in cities with high public awareness of air pollution and air pollution data, it has even been argued that neighbourhood air quality readings are starting to directly impact house prices. In the US, this also leads to a racial divide. A study in 2006 found that reducing NO_2 levels experienced by ethnic minorities down to levels experienced by whites would reduce ischaemic heart disease mortality by around 7,000 deaths per year – a similar impact, said the paper, as three million adults quitting cigarettes. This also means that targeting the areas with the highest concentrations of pollution can have a greater overall health impact, and help the most vulnerable members of communities, in comparison to a blanket PM2.5 or NO_2 target that covers a whole city, or even a whole country. 'We have areas where many parts of a region meet the federal standards, but there will be pockets of pollution that are well above admissable levels,' says Mary. 'So, the purpose of environmental justice is not just to meet a standard, but actually benefit public health. You want to start focusing your attention on how to achieve the best results for everybody. This is one of the shifts in thinking that is going on now.' For example, new school sites now cannot be within 500 feet (152 metres) of urban highways.

A 2017 paper on southern California in the journal *Environmental Research Letters* states that community-targeted emission reductions such as low-emission zones (LEZs) and truck re-routing 'can reduce environmental injustice while also meeting multiple other air quality management goals'. It again found that mean exposure to PM2.5 from diesel engines was 38 per cent higher for ethnic minorities than for ethnic whites.[1] This helps to explain Mary's reluctance to move from a PM2.5 regulatory standard to a nanoparticles standard: community-based interventions based on environmental justice and levels of exposure (or particulate number) are potentially the best ways of bringing down overall PM2.5 levels too. As Suzanne Paulson at UCLA also puts it, 'there doesn't seem to be much question that if you lower the PM2.5 things get better. So as a standard it's OK really ... emissions controls for vehicles are focused on PM2.5 and NOx and CO, but the controls for those have led to a marked drop for ultrafine particle emissions as an unintended consequence.'

The environmental justice movement also answers another issue, which I encountered in every city I've visited or researched: why should poorer people pay more in tax to help affluent people to buy a nice new electric car? I put this to Mary. 'People living in the communities that suffer the most from air pollution, in general, are most supportive of action being taken to address their problem,' she responds. 'First of all, if I am driving my electric car or zero emissions vehicle through a poor community, or I'm a commercial operator that runs trucks through these communities, I am impacting their health – it is not just the health of the person who buys the vehicle – everybody around them is exposed to their pollution ... people who have the worst pollution in their neighbourhood are saying, this isn't acceptable and you have to clean up to address our health problems ... the most dramatic illustration of how this can work is in our greenhouse gas reduction, where we direct the proceeds of our cap-and-trade programme with a legal requirement that a third of all the funds that are collected by the state from the auctioning of allowances will be diverted to the communities that have the worst air quality.'

The 'cap-and-trade' scheme was signed into state legislature by Governor Schwarzenegger in 2006. With just 450 businesses responsible for about 85 per cent of California's total greenhouse gas emissions, from 2013 they were given a maximum cap for the amount of pollutants they could emit, starting at 90 per cent of average emissions and declining by 3 per cent annually from 2015 to 2020. If they emit above the cap, they have to buy an 'allowance'. This differs from a simple system of fines, in that these allowances are tradable, and their price goes up over time, as does the number of regulated companies. The quarterly auction of allowances makes the state a lot of money and reduces emissions. Proceeds of around $900 million a year have been spent on a range of clean air initiatives such as creating parks, planting trees, making low-income family homes energy efficient, grants for electric cars, installing EV charging stations and high-speed rail; and, as Mary said, a quarter of all the money generated by the state's cap-and-trade is ring-fenced for environmental justice programmes within disadvantaged and low-income communities.

However, California doesn't have the power to regulate all its emissions. Interstate or international transport such as road haulage, trains, shipping and air travel do not come under state control. And the EPA under the Trump administration is going in precisely the opposite direction to California.

In February 2016, the Attorney General of Oklahoma, Scott Pruitt, was a political unknown on the national stage. He was just doing what he loved best, fighting the EPA in court on behalf of his oil and gas buddies. Since taking office in 2010, Pruitt had sued the EPA on no less than 14 previous occasions, each time to relax emissions laws. His early suits included trying to roll back the Mercury and Air Toxics Standards, which limited the amount of mercury, arsenic and other toxic pollution from power plants. It was put in place by the EPA to prevent up to 11,000 premature deaths and 130,000 asthma attacks. According to an investigation by the *New York Times*, companies or trade associations in 13 of these cases were also financial contributors to Pruitt's political causes. In December 2016, the same Scott Pruitt was appointed by the newly elected

Trump administration to be the head of the EPA. The fox was now in charge of the chicken coop. The budget for the federal department responsible for enforcing the Clean Air Act was cut by 31 per cent and its workforce by 21 per cent. Specific cuts included the EPA's federal vehicle and fuel standards and certification programmes and grants to help support state and local air quality programmes.

One year on from Scott Pruitt's swearing-in ceremony, I asked Mary if CARB have felt the impact from the changes to the EPA. 'Well,' she sighs, 'there's a general direction of moving backwards or stopping progress on all kinds of environmental regulations, and certainly air is no exception.' I ask if the Californian Clean Air Act waiver effectively protects her jurisdiction, in regard to automobile emissions? 'It's not a blanket waiver that says "you can do whatever you want to, California", it is case by case, regulation by regulation.' Historically, out of hundreds of waivers sought, the EPA has only ever denied California one, during the Bush administration, and even then only briefly. At the time of writing California was yet to submit a waiver request to the Trump-era EPA. 'We have to try, and we have to presume that the government will follow the law,' says Mary. 'If we can demonstrate the need for tighter emissions, and if there's a demonstration that there is alternative technology that can feasibly achieve that, then we are entitled to the waiver.' James Thornton recalls a similar situation during the Reagan years, when Anne Gorsuch was tasked by the President to 'bring the EPA to its knees', but ultimately failed. Whether Pruitt could succeed was 'an open question. You can't get rid of the laws, even with this Congress you're not going to be able to repeal a Clean Air Act or a Clean Water Act.' We got the answer to whether Pruitt would succeed in July 2018, when he became the latest in an already long list of senior Trump appointees to resign amid scandal allegations. His immediate successor, however, hardly suggested a change in direction: Andrew Wheeler was a former lobbyist for Murray Energy, the coal mining company.

A lot of hope now rests with California. Since the Trump presidency, Californian Governor Jerry Brown has arguably ratcheted up his environmental commitments. Mary Nichols confirms that Brown is pushing for a UK-style commitment to ban diesel and petrol cars. He has extended the state's cap-and-trade program through to 2030, the same year that the state must hit a 50 per cent renewable energy target. Brown has since announced a 100 per cent renewable target by 2045 and called for the whole United States to go to 100 per cent by 2050. Meanwhile, under the Clean Air Act any other state can still choose to adopt California's stringent vehicle emissions standards. Fourteen states currently do so, including Pennsylvania, Connecticut, North Carolina and New York City – referred to as the 'CARB states'. This means that California's rules cover 135 million people, more than 40 per cent of the US population. And if the EPA continues its decline, more blue states may join the list. California's 'Pied Piper of Clean Air' role is now arguably more important than ever.

Paris: A journey without cars

I'm standing in the middle of the Place de la Bastille with only nine minutes to go before the '*Journée sans voiture*' (Car Free Day). It is 10.51 a.m., and the car ban officially starts at 11 a.m. But on this grey, cold and drizzly Sunday in October, there are still a surprising number of cars scurrying around. They surely won't have long enough to make it out of the central zone before the car ban kicks in? I check my watch. Now they only have eight minutes. I'm both waiting to see what happens at 11 a.m., on what is normally one of Paris's busiest thoroughfares, and then to meet Charlotte, an air pollution analyst from the Airparif team. She and her colleague are due to arrive by bicycle, and I'm hoping they will be easy to spot.

The city's first Car Free Day in 2015 saw all but public transport and emergency vehicles banned from the roads, resulting in NO_2 levels plummeting by 40 per cent in parts of the city. The success was such that Mayor Anne Hidalgo posted on Twitter: 'We might envisage days without cars more often … perhaps even once a month.' Several streets in central Paris subsequently vowed to remain car free on the first Sunday of every month, including the Champs Elysées. The *Journée sans voiture* was also retained as an annual day, but with bolder ambitions. While the first two years were limited to the city centre, the 2017 day, when I arrive, is set to take effect across the whole of Paris. Christophe Najdovski, the deputy mayor in charge of transport, told the newspaper *Le Parisien*: 'The idea … is to show that you can live in the city without having a car … and will allow everyone to rediscover a quieter and less polluted city.' This was something I had to experience.

It is now 11.10 a.m., and there are still cars on the road, but I have seen some officials in yellow high-vis *Journée sans voiture* vests stop and remonstrate with a couple of vans. And I needn't have worried about whether I would spot Charlotte. Also dressed in high-vis, she arrives on a bike pulling a large trailer behind her the size of a mobile coffee shop, covered in blue tarpaulin and plastered with Airparif logos and a giant pollution map of Paris. She lifts the tarpaulin to show me the impressive amount of kit that has been packed into it, monitoring both NO_2 and PM2.5, with a tube poking out the top to suck in the air. Her plan is to cycle it around central Paris all day, taking readings to see the difference in pollution levels compared to non-car-free days. She tells me that the pollution-mobile (my words, not hers) has been used before to show the impact of segregated cycle lanes, and found that the pollution levels were 30 per cent less in a cycle lane compared to cycling on the road among the cars.

After Charlotte cycles off, I plan to rent a Velib – a rented bicycle from the city's bike hire scheme, and one of the first in Europe when it was launched in 2007 – to set out to meet the people triumphantly reclaiming the streets by bike or on foot. Before the final cars hurry away I take my Egg out and it takes readings in the single figures. A blustery day like today with drizzle blowing through the air is good conditions for clear air anyway. I was expecting the Velibs to be like gold dust today, but the racks at the Place de la Bastille are surprisingly full. I haven't rented one before, and stare blankly at the instructions on the machine. It's not nearly as user-friendly as other city bike schemes I have used. I have a ticket with a lot of numbers on it and a personal four-digit code, and the bikes themselves also have numbers. Completely confused as to what number I now need to enter or in which order, after two failed attempts I realise that a gaggle of press photographers have appeared around me taking photos. Needless to say, this isn't helping. I get increasingly flustered, now just punching numbers in at random, doing my best exasperated Gallic shrug, and wondering how to exit from this situation with a shred of dignity. Another Velib cyclist rides towards me and dismounts. I look up to ask her pleadingly for help with this damn machine, but before I get a chance the photographers shove me out of the way and surround her. I realise it is Anne Hidalgo, the mayor of Paris, arriving in La Bastille to mark the start of Car Free Day with a photo shoot. In person, she is Parisian from head-to-toe, with an ability to look elegant even on a bicycle while wearing a bike helmet. Before I'm able to blurt out a better question than how to use the damn Velib machine (although that might not have been such a bad question), she is immediately surrounded by press, security and a TV crew. The PR manager glares at me as he whisks her away.

I return to the Velibs, still none the wiser. I ask the next person – this time not the mayor, but an ordinary guy, and after he gestures with authentic Gallic shrugs and an 'et voilà!', I gratefully pull a bike from its rack and start pedalling, eager to experience what a major city is like without cars.

Rather than 'Car Free Day', however, it is quickly apparent that it is 'Fewer Cars Day'. Buses, taxis and private car-hailing services such as Uber still ply the streets, and potentially in greater numbers than usual as, being the only four-wheeled game in town, they can charge a premium. I was expecting streets reclaimed by celebratory cyclists and pedestrians, walking arm-in-arm down the tarmac.* However, pedestrians are still largely restricted to the pavements, and cycling – while less nerve-racking than on a usual Parisian day – is still the familiar urban exercise of truck and car avoidance. My first Velib experience also continues as smoothly as it started: my first bike has a puncture, and I have to return it. But finally, I do start to encounter the joyous reclamations of the streets that I had hoped to see. Whole families cycle side by side on roads that they wouldn't typically dare go down; environmental protesters begin to turn up on bikes they've turned into 'cardboard cars' as an indicator of how much space cars needlessly take up. Most striking is a group of up to 100 gyroscope riders – something I hadn't encountered before, like motorised unicycles but without the seat, which appeared to have spawned a counter-culture among kids and men dressed largely as gamers, many with customised helmets. One rolls a cigarette in his hands as he travels at speed, barely looking ahead. They carry a 'fuck you' attitude that I wholly approve of. The anti-car movement needs some angry, 'fuck you' types. And Paris has never been short of those.

I arrive on my Velib in Place de Stalingrad for lunch, hoping to see a large procession that I've heard is to end there and gather for a 'waste-free picnic'. I arrive before the procession and find a celebration is already under way. A large stage has been set up with a rapper, then a *chanteur* (I later learn they are HK and Mali Karma), and the Car Free Day organisers are dragged up on stage for lots of embarrassing dancing, and to sing a number of environmental protest songs. When the procession arrives, it is made up of ordinary

---

* Something I experienced in Bogotá, Colombia, where around 100km (60 miles) of roads are closed each Sunday, including la Séptima, the major road into the city centre. Each Sunday morning, it is full of cyclists, pedestrians, roller-skaters, and not a single car in sight. Many cities from Cape Town to Ottawa have since copied the idea. But it's not every road; there are still parallel roads full of cars. Paris's big idea was to make it apply to every road, for one day only.

cyclists, several of the 'cardboard cars' and even, for reasons I never quite get to the bottom of, a cardboard lighthouse about five metres high, manned by a man dressed as a ship's captain, smoking a pipe and clacking seashells together to the rhythm of the music. I join in the waste-free lunch (a chickpea stew that tasted similar to the cardboard it was served on) and find myself drawn into the celebratory spirit. This is a recognition of how significant a step a day like today can be: that people could reclaim the streets from cars, which could reclaim the air from traffic emissions, too.

I ring Mariella Eripret, one of the Paris *sans voiture* citizen group organisers, whom I have arranged to meet. We can barely hear each other over the music. I wonder if she is one of the organisers with green T-shirts and red faces dancing on stage, but fortunately, she isn't. When we talk, she fills me in on the background. The idea for Paris *sans voiture* came from a group of citizens, not from City Hall. 'As a Parisian cyclist, I am very annoyed not only by air pollution but also noise pollution and space taken by cars in the public space,' she tells me. 'We were a few people who met during a festival we organised (*Festival des utopies concrètes*), and we imagined or dreamed about a real and complete car-free day in Paris. At the beginning, I have to admit that I did not believe it could happen ... The mayor took a while to answer. But in the meantime, we got in touch with her deputy mayor [Christophe Najdovski] who totally supported us. Anne Hidalgo was supposed to announce the car-free day on 7 January 2015 but she finally did not because of the terrorist attack at *Charlie Hebdo*. We wrote her another letter saying we should not abandon the idea, even in that terrorist context, and that we needed such events more than ever ... We want this day to be an opportunity to see and live the city differently, to make people aware of the space taken by cars on the public space, and the pollution they cause.' And what's the cardboard lighthouse all about, I ask? 'I don't know exactly ... People were just invited to join the procession riding any kind of engine, as long as it had no motor!'

I end the day walking along the Right Bank of the Seine – until last year, one of the busiest, most polluting roads in the city. Last year, Mayor Hidalgo made the decision to close it indefinitely and open it up to walkers and cyclists. Here the celebration continues and this time it is permanent. Children play on climbing walls and wooden obstacle courses. Where there were once lorries belching fumes there are now trees in pots. Even when the drizzle eventually gives way to full-blown rain, there is a tangible happiness on the Right Bank. No one needs a 'fuck you' attitude down here, because the cars have been banished for good, with no exemptions. Every day is a '*journée sans voiture*'. And boy, it feels good. The air feels clean. Airparif studied an earlier partial pedestrianisation of the Left Bank of the Seine and found that NO_2 levels by that section of the river had dropped by 25 per cent. The NO_2 that remains comes from the road a few metres above, where cars can still drive. However, even on the upper road, NO_2 was found to be 1–5 per cent lower than before, because there was no longer the combined effect of NO_2 mixing in from below.

Tyler Knowlton, a Canadian living in Paris, who works for the air monitor start-up Plume, tells me that further along from the fully closed section of the Right Bank, two lanes of traffic have also been narrowed to one, and 'there is a double bike lane there now. It's like "everyone be damned, this is for bikes now" – that's incredible. My stress level went down by like 20 per cent trying to get to work in the morning ... at the weekend in the 15th [*arrondissement*, or district] on the Left Bank, I used to have to chain my bike to a street sign. Now there is all new bike parking, it is repaved, they put a dedicated bike lane in. All over the city there is all this new cycling infrastructure.' The city plans to double the number of bike lanes from 700km (434 miles) in 2015 to 1,400km (870 miles) by 2020, including segregated cycle lanes wherever possible.

Just as the growth of cycling blossomed following the introduction of the Velib, the city hoped to do the same with electric cars. Autolib, the city's electric car-sharing service, with a docking station model like the Velib, was launched in 2011. A fleet of over 4,000 electric cars and 6,000 dedicated charging points and docking stations were prominently positioned across the whole Paris region. The vehicles were available for short-term rental to members of the scheme, exactly like the Velib bikes. On average, each Autolib vehicle was found to replace the need for three private cars – by 2016, it broke above 100,000 annual registered subscribers. One of the key attractions was a guaranteed parking space at the start and end of your journey. Being Paris, where parking is something of a contact sport, many of the grey Autolib cars were covered in bumps and scrapes; far from being some green vanity scheme, these were lived-in cars that carried the scars of regular use.

Autolib came to an end in July 2018 when a budget shortfall led to a falling out between the city authorities and the private company contracted to run it, the Bolloré Group. However, by that point, arguably it had already proven the concept – electric, shared vehicles are very popular amongst urbanites. Autolib soon had competition, too. In June 2016, a fleet of 1,600 'Cityscoot' electric scooters appeared on the streets of Paris. Unlike the Autolibs, these could be left to park anywhere within the central zone – you use an app map to find the nearest one, then hop on. Somewhat worryingly, users don't even need a driving licence. Nor do they need to recharge the scooter after use. 'I was a beta tester last year,' enthuses Knowlton. 'You can leave it anywhere, just get on. They have a bunch of people that go around with flat-bed trucks and pick up the ones that need recharging. I've never not been able to get one.'*

All diesel cars will be banned in Paris from 2025. Given that in 2017, almost half of all cars were still diesel, it sounds

---

* The popularity of 'dockless' bikes also poses a threat to Velib, and dockless bike operators such as Ofo and Mobike make picking up a bike extremely easy and user friendly. Unfortunately, they also make abandoning a bike almost anywhere very easy, too.

unachievable, but the pioneering Autolib scheme and the imitators it spawned give us a glimpse of what that future might look like. There is also a clear, graded pathway in place from the city authorities to phase out diesel. As of January 2017, the oldest, most polluting diesel cars were banned from the city's streets during the day. All cars, motorbikes and lorries in Paris must now display a coloured, numbered disc called Crit'Air, based on the age of the car and its emissions, ranging from Crit'Air 1 (electric and hydrogen-powered vehicles) down to Crit'Air 6 (older, mostly diesel, vehicles). From July 2016, cars registered before 1997, deemed the cut-off point for Crit'Air 6, were prohibited from entering Paris on weekdays between 8 a.m. and 8 p.m. The carrot offered to those owners getting rid of their cars have included discounts for Autolib, a year of free public transport, financial assistance for businesses, free Velib membership, and even €400 to buy a bicycle. From 1 July 2017, the ban was extended to vehicles bearing the Crit'Air 5 disc, which included diesel vehicles registered before 2001. And so on, each July, until we reach zero diesel cars by 2020. The Crit'Air scheme also works as a viable alternative to the emergency odd/even number-plate scheme used in Paris and many other world cities during peak episodes of heavy smog. Rather than arbitrarily taking half the cars off the road, irrespective of how polluting they are, now the Parisian authorities can implement, for example, such measures as 'only cars displaying Crit'Air 1 & 2 certificates will be allowed into Paris, until the smog clears'.

The day after *Journée sans voiture*, Paris is back to normal, and full of cars. At the Airparif offices, I ask Amélie Fritz if they have studied or modelled the impact of Crit'Air? 'The what?' she asks, confused. I repeat it. 'Ah, *Crit'Air*,' she says, correcting my terrible French accent. She clicks through her computer to find some files. The first restrictions put in place in 2016 affected 'around 2 per cent of the car fleet', she tells me. But by removing this 2 per cent, it reduced air pollution levels by '5 per cent of NO_2, 3 per cent of PM10 and 4 per cent of PM2.5. The second step from July 2017, Crit'Air 5, is only 3 per cent of vehicles … This 3 per cent of vehicles will reduce NOx by 15 per cent, and 11 per cent of PM2.5. It is', she says, with a large degree of understatement, 'potentially quite a good measure.'

That said, Europe has had a road map for reduced emissions before, with the Euro 1–6 car regulations. As the fallout from the VW scandal showed, those milestones were never met on the road. And politicians come and go. There are powerful motorist lobbies in Paris, as in any other city, calling for a repeal of the cycling lanes and road closures. According to the latest Airparif report, more than 1.4 million Parisians are still exposed to levels of pollution that do not comply with the regulations for NO_2, with 'the health of the inhabitants of Paris living along the traffic routes and in the heart of Paris … the most affected'. Paris, perhaps more so than anywhere else, left me with a strong sense of our two possible futures hanging in the balance: it offers glimpses of what zero emissions could look like, but remains largely stuck in the stranglehold of fossil fuel emissions.

The global awakening

Globally, we face the same enemy. The European Union has set a 2050 target of reducing emissions from the transport sector by 95 per cent. To deliver this it means that almost every car, van, bus and lorry on the streets of European cities needs to be zero emissions by 2050. Currently, the only zero emissions vehicles are battery or hydrogen powered. Given the average age of vehicles on the road is 15 years, it also effectively means that no diesel or petrol vehicle can be sold after 2035. Several cities are leading this drive, with Oslo aiming to provide 100 per cent renewable energy-powered public transport by 2020, and Amsterdam by 2025. In October 2017, the mayors of London, Paris, Los Angeles, Copenhagen, Barcelona, Quito, Vancouver, Mexico City, Milan, Seattle, Auckland and Cape Town committed their cities to procure only zero emission buses from 2025 and ensure that a major area of their city is zero emission by 2030. There are plans to ban both petrol and diesel cars in India by 2030.

In Germany, where ClientEarth are very active, James Thornton tells me 'we have had ten air quality cases, and we have won all of them at my last count … courts in Stuttgart, the heart of the motor industry, and in Munich, have said that they will ban diesel by court order … judicial behaviour is changing because they understand there is a public health emergency.' Even in the European capital of coal, Poland, Maciej Rys, the founder of Krakow-based Smogathon, tells me, 'the government in Poland took some serious steps … From the beginning of 2018 you cannot sell these old tech chimneys or furnaces for your home – you cannot sell them or install them … I think there is a big revolution coming in our whole approach, especially to coal. You see even coal miners saying we know this is going to end, even they are convinced that this is not the future.'

We have reached the limit of what can be achieved by incremental improvements in engine technology and combustion efficiency. The Euro 6 standards were the end of that line, and real-world tests show that the car manufacturers just can't meet them. In October 2017, just before my second daughter was born in an NHS hospital in Oxfordshire, Oxford council announced plans to introduce what has been described as the world's first 'Zero Emission Zone' in Oxford city centre. The proposal would see diesel and petrol vehicles banned from the city centre in phases, starting with a small number of streets in 2020, potentially moving to the whole city centre by 2035. Oxford city Councillor John Tanner said that a 'step change' is now needed: 'All of us who drive or use petrol or diesel vehicles through Oxford are contributing to the city's toxic air.' Everything is pointing in the same direction: the end of combustion for energy and transport, and the rise of electrification.

CHAPTER EIGHT

Electric Dreams

Milton Keynes used to be, and perhaps still is, a bit of a joke in the UK. A modern city built around an American-style grid system in the 1970s, it has never been fully accepted as English (where roads should be narrow and unnavigable). The new city looks and feels like a huge business park and introduced the nation to the much-loathed, albeit rapidly copied, roundabout. Visiting it in July 1974, the *Daily Telegraph*'s Christopher Booker, presumably with a clothes peg on his nose, decried it as 'the utterly depersonalised nightmare which haunted Aldous Huxley in *Brave New World* just forty short years ago'. All of which goes some way to explaining why I'd never been to Milton Keynes before, despite living barely 30 miles away from it for many years. But Booker, suffering from the sensory under-load of mono-tone concrete greys, misread the hippy intentions of its chief planner Richard Llewelyn-Davies, who was attempting something altogether more spaced-out. 'The future is rather indeterminate,' he told the *Illustrated London News* in 1970. 'In planning of this sort it's futile to make guesses. You have to design a city with as much freedom and looseness of texture as possible. Don't tie people up in knots, man.' OK, I added the 'man', but the rest is in his words.

For Llewelyn-Davies the skeleton of the city, the infra-structure, was more important than the buildings in isolation. Much like the Pompidou Centre being built at the same time in Paris by the exciting young architects Renzo Piano and Richard Rogers, which externalised its interior workings and made the functional beautiful, so the design of Milton Keynes was to do the same for town planning. According to the CMK Alliance, a group of council and local business leaders, 'The infrastructure was to be the eternal skeleton, muscles, arteries and nervous system of the entire urban body, bringing it to life.' Milton Keynes was one of the very first city centres designed to be pram- and wheelchair-friendly, 'providing "barrier-free" access for all, using underpasses and kerbs at a level with pavements in parking areas and along slow streets … The public realm is its greatest achievement, providing a framework in which the buildings and activities of the city centre might come and go over time.' Continuous ground-level footpaths connect surrounding estates; delivery trucks have separate service bays away from the main roads; pedestrians and cyclists are carefully removed from traffic on 230km (140 miles) of segregated lanes (before anyone had even used the term 'segregated lanes') called Redways, largely below road level, using a network of shallow underpasses to dip under the road intersections.

The master plan, with wide boulevards and generous tree planting, also lends itself to clean air. Unlike the narrow lanes lined with tall buildings of older cities, Milton Keynes's streets avoid the 'street canyon' effect, with lower blocks, much more widely spaced, meaning that pollution doesn't get trapped. It was always planned as a green city, with more than 20 million trees and shrubs, 15 lakes and 11 miles of canal and riverside. More than 40 per cent of the urban area is green space. Perhaps unsurprisingly, the people who live there really like it. Milton Keynes is one of the fastest-growing cities in the UK, increasing by 16 per cent between 2004 and 2013 to just over a quarter of a million people, and its population is projected to double by 2050. Currently, there is a heavy reliance on cars, with above national average car journeys to work (61.8 per cent) and above average car ownership (83 per cent). But the city is doing more than anywhere else I visited for this book – including Paris and its AutoLib project – to get everyone into electric cars and electric transport. Its many UK firsts include being the first city to trial autonomous driverless pods on public thoroughfares; opening an electric vehicle (EV) showroom in a public shopping mall; installing over 200 on-street EV charging points; installing more than 50 rapid chargers within the city limits; and introducing wireless charging for electric buses.

Walking through the city I pass more electric vehicles and electric vehicle charging points (and electric vehicles charging at electric vehicle charging points) than I have seen anywhere before. At the electric car showroom in the shopping centre – nestled between Boots the chemist and F. Hinds Jewellers – there are three electric or plug-in hybrid models on display. I am also here to meet the man responsible for electrifying Milton Keynes: Brian Matthews, head of transport innovation at the local council. We go to a meeting room where a photographic history of the evolution of the electric car runs chronologically along the walls. An early example from the 1970s looks like a yellow triangular doorstop on wheels – the 1974 CitiCar, produced in Florida in response to the 1970s fuel crisis. It had a range of 40 miles, which wouldn't get you very far in Florida. Milton Keynes installed 200 EV charging points before there were any EVs on the road, so my first question is 'why?' 'Because the evidence was growing,' explains Brian. 'Nissan had launched their car [the Nissan Leaf – the UK's first mass-produced electric car, built in Sunderland] ... so we needed to be prepared and be ahead of this. But we also had 25,000 parking spaces, so even putting 200 in just for electric vehicles was a drop in the ocean.' Now, he says, 'we are setting the model for how electric transport could be integrated into a modern city.'

After we leave the car showroom, Brian takes me on a tour of Milton Keynes – first on foot, then in his plug-in hybrid electric car. We pass a parking space that says 'premium rate £2 an hour, free for EV'. Brian's car is plugged in to a charger in a public EV parking space. He swipes a card across the charging point, it recognises him as the owner and releases his charging cable, ready to put back into the boot. It also tells him how much charge is left in his battery. Once safely out of the wind and into the front seat, he tells me, 'My commute is just under 30 miles, so I can do my whole journey on one charge here, and charge at work for the way home.' He points at a rapid charging point as we drive past – 'that will charge a vehicle in 20 minutes to 80 per cent. The one I was plugged in takes two hours for my battery.' His charging-scheme membership card is with Chargemaster, using energy supplied from renewable sources.

'We're going to go and look at the electric bus project,' he tells me, as we leave the central business district and approach one of the myriad residential cul-de-sacs that radiate out from it. 'We run a full-electric bus for 15 miles through the city, it's one of our longest routes.' Unlike almost any other EV, however, this one never needs to stop and plug in during a full working day. 'How it operates, as we'll see it in a minute, is it reaches its turnaround point at the end of the route and dwells over the charge unit … Oh, it's just arriving – that's the bus!' He pulls in by the side of the road, and we jump out to watch the bus as it stops. A single-decker (the 'number 7, via Stanton'), it looks no different from any normal bus in any UK city until you notice the battery pack that sits on top of the driver's end, like a little hat. As we watch, a metal plate lowers from underneath the bus and settles onto metal plates buried into the road asphalt. 'It is picking up an opportunity charge for five to seven minutes, which means the battery is topped up and then it can start its route again', explains Brian, giving a running commentary. 'It has a full charge overnight, it goes out full, and then as it goes about its route it picks up enough charge like this at both ends to keep it going all day. You don't have to take it off the road at any point. A driver changeover happens on normal routes at turnaround points, so this is no slower than a normal bus route.' How exactly is it charging? 'It's like a toothbrush in a toothbrush holder – it's the same technology, only bigger,' he explains. 'There's no external charge risk, you could put your hand on it. We've had to prove it with pacemakers, everything, safety is not an issue.' How long have you had this in operation? 'It's been over three years now.'

While many cities wring their hands over how to make buses fully electric, a normal, in-service electric bus has been running for three years already – and it's in Milton Keynes?!

Brian tells me that passenger numbers on this route are up 3 per cent per annum, despite nothing else having changed about the route, and the drivers prefer the smoother, quieter ride, too. The next stage of the plan is to make all bus routes in the city electric. 'You can't afford rail or trams in small cities. And there's no need now. This is fit for purpose,' says Brian. I ask if the same wireless charging plate technology could eventually be utilised for electric cars, too. 'That's what we're looking at … Initially we're thinking in taxi ranks. If you get the power ratio up, you could have them in bus lanes, even as you wait at a roundabout. The fascinating thing about EV is, there is the potential of looking back at the anxiety of internal combustion engine [petrol and diesel] cars and wondering where the next petrol station is, as the real "range anxiety" – if you had multiple top-up charging points everywhere for EV, you'd never have to worry about the next charge.'

There are other variations on this theme emerging globally, too. In Jinan, China, the world's first 1km (1,100yd) stretch of solar motorway opened over Christmas, 2017, with transparent concrete laid on top of photovoltaic panels producing enough electricity to power the street lights and, in theory, add to the charge of the cars driving over it. Stanford University brought that theory a step closer to reality that same year, with the first team to successfully transfer electricity wirelessly to a moving object.[*] If this and the Chinese solar road surface ever managed to join forces, it would in effect herald the age of range-less road transport. Meanwhile, back in today's world, even the rapid charging points still take 20 minutes to 'fill' your EV car. 'Maybe there is some behavioural change needed with 20-minute charging stops,' admits Brian, as we drive quietly along, 'though that is potentially not a bad thing if

---

[*] Although admittedly, the Stanford team's breakthrough was only able to wirelessly transfer one milliwatt of power, which would theoretically take 5,993 years to fully charge a 40kWh Nissan Leaf. My thanks to my engineer brother-in-law Ewan Jones for working that one out!

it breaks up fatigue in driving. But technology development is such that soon that 20 minutes [rapid] charge will be 10 minutes, then 5 minutes.'

The air quality credentials of electric vehicles over internal combustion engine (ICE) vehicles are obvious: there are no exhaust emissions. Zero. No NOx, no black carbon, no combustion-derived nanoparticles. There is still PM2.5 from the resuspension of road particles, but with no ICE cars on the road soon those resuspended particles would not contain black carbon or combustion-derived particles. Even metals are reduced as you use your brakes far less – as you release your foot from the accelerator, the car immediately slows down. The typical braking system of electric cars is also regenerative, capturing and reusing the energy produced by braking. Dr Beevers at King's has modelled what an all-electric fleet would look like in London and informs me that NO_2 effectively disappears, while traffic-derived PM2.5 reduces by 50 per cent. The modelling hasn't yet been done for nanoparticles, but given everything we know about their formation and proximity to source, I'm going to stick my neck out and say they effectively disappear too, meaning that total particle number by roadsides would be far less than 50 per cent. Even in the very worst-case scenario, if all the electricity to run EVs came from coal-fired power stations, then as long as the power station is far removed from human habitation, the urban exposure to combustion particles is still vastly reduced. But in reality, that worst-case scenario won't happen, and already doesn't exist. Coal is rapidly on its way out. Power generated by wind, solar and biomass in the European Union overtook coal for the first time in 2017.

The lifetime emissions from electric vehicles versus ICE vehicles vary from study to study, but they consistently show electric vehicles coming out on top. A two-year-long study by the US science body UCS (Union of Concerned Scientists), for example, found that EVs 'produce less than half the global warming emissions of comparable gasoline-powered vehicles, even when the higher emissions associated with ... manufacturing are taken into consideration.' After 6 to 16 months of driving, your EV is increasingly cleaner than ICE from that point on. ICE cars have a 30 per cent energy efficiency rate, losing most to heat, while car batteries typically have an 80 per cent efficiency rate. Plus, the manufacture of EVs is far simpler than ICE cars – there are far fewer parts, making mass production quicker and more straightforward, and reducing lifetime maintenance needs. In terms of healthcare costs, a joint study by the University of Oxford and University of Bath found the health damage from diesel vehicle emissions to be five times higher than the exhaust fumes of petrol vehicles and 20 times greater than the dust kicked up by electric vehicles (in other words, all vehicles resuspend road dust, but only ICE vehicles add toxic combustion particles and NOx into the mix).

Replacing petrol and diesel with electric batteries is part of the blueprint to become a clean air city. But while my visit to Milton Keynes filled me with a lot of optimism, it was mixed with frustration. If this is all available, proven, existing technology, then – given the obvious clean air benefits – why isn't electrification happening everywhere?

★ ★ ★

On a Wednesday morning in September, I return to the Millbrook testing ground for my second visit. This time for the annual 'Low Carbon Vehicles (LCV) show', the motor show for petrol-heads who no longer dig petrol. I drive there – because it is pretty much the only option – and on arrival I join a long group of cars queuing down the 'straight mile' of the test track waiting to be shown to a parking space. Every car I see has only one person inside, the driver. Most people arrive late due to congestion on the motorway. Giving the show's introductory speech is Brendan Connor, chairman of Cenex (who describe themselves as the 'independent, not-for-profit, low carbon technology experts'). It's his last

LCV show before stepping down as chair. The first LCV show 10 years ago was, he reminisces, a windswept event that only lasted a handful of hours 'before we all gave up and went home'. There were only 10 vehicles on display. 'And my recollection is that only two of them actually worked. And the two that worked technically qualified as "quadricycles".' Now, 10 years on, there are over 130 LCV vehicles on display, and over 230 exhibitors. His message is clear: low emissions vehicles have gone from a niche tech eccentricity to mainstream within a decade. All the major car companies are here, showing off their latest electric models.

Also at the LCV show, Graham Hoare, a director at Ford Motor Company, confirmed that: 'the electrification dream is actually going to be an electrification reality very quickly, surprisingly quickly, in my opinion. And we're going to see a complete change in the way vehicles are used and owned. And all of this is going to come together at a much faster rate than I think all of us predicted.' I also see Nobusuke Tokura, senior vice president, Nissan Technical Centre Europe, who paints an ambitious and exciting picture of the future. In 2009, there were only six models of EV available globally. By 2011 (when the Nissan Leaf was first launched), there were around 36. As of March 2017, Tokura informs us, there were over 139 models and a cumulative 1.28 million sold globally. The trajectory is doubling year on year. The Leaf alone sold 283,000 worldwide by 2017, making it then – pre-Tesla Model 3 – the world's number one electric car by sales. 'When we launched the Nissan Leaf, everybody was laughing about Nissan,' he cheerfully admits, 'it was too early, too ridiculous to launch a product like that. But … it was a very good success. The Nissan Leaf achieved so many things.' This includes, he says, 3.5 billion emissions-free kilometres (2 billion miles) driven on the road. 'I am really proud about that result,' he says, beaming. The first Leaf had a range of 109 miles on a single charge. By 2013 this was pushed up to 120 miles. In 2015, it went up again to 156 miles. In late 2017 the Nissan Leaf 2 came out with a range beyond 200 miles. The range of the Nissan Leaf – just one model of

EV – almost doubled in just six years. Given that the average ICE car can typically travel around 300 to 400 miles on a full tank of petrol, it's not hard to imagine this will soon be matched and exceeded by EV* – with the advantage of having an equivalent 'gas station' at home in the form of plug sockets.

Tokura plays a video of Nissan's vision for the future city. Nissan believes that streets will be peppered with the same wireless charging pads I saw in Milton Keynes. This isn't just to charge the car, however. 'Come the morning, your house and the grid can draw energy straight from your car, powering your home as you start the day,' says the video narrator. This would in effect solve the energy storage question posed by renewable energy: how to store all that excess energy from wind on windy days or solar on sunny days, to draw on during the windless, overcast days? The answer, says Nissan and many other EV advocates, is EV batteries. In the 'smart street' of the near future, says Nissan, 'the cars, road, houses and grid are all in sync and connected. Recycled [EV] batteries can be used for smart home storage, so that no clean energy goes to waste … space once occupied by car parks and fuel stations could be replaced by green spaces, building a cleaner, kinder environment for our children.' The car, home, street and battery in effect become the refuelling station.

I see other, nearer-term visions of the 'electric street' at the LCV show too: two separate charging companies at the exhibition are looking to partner with local authorities to fit EV chargers onto existing street lamps, tapping into the mains without diminishing the light – some street lamps have already been retrofitted this way in London, with more to follow in Oxford. Suddenly there is the possibility of charging stations on every street corner without even having to build any extra infrastructure.

---

* Already in 2018, ICE vehicles were very close to being superseded. The Tesla Model S will be available with a long-range extender capable of 335 miles (539km).

Volkswagen, thanks in large part to their public fall from grace, has also seen the light when it comes to electric vehicles. The VW Group is planning to have 50 models of fully electric cars and 30 plug-in hybrid models on the market by 2025. In 2017, it already had the e-Golf and smaller e-up!* as well as electric/petrol hybrids including the Golf GTE and Passat GTE. VW also announced that there will be 'at least one electrified version of each of the 300 or so Group models across all brands and markets' by 2030. The iconic VW campervan, much loved by surfers and festival-goers, has been reborn as the electric VW Buzz. Matthias Müller, chairman of VW, has stated unambiguously: 'We have got the message and we will deliver. This is not some vague declaration of intent … The transformation in our industry is unstoppable. And we will lead that transformation.' By December 2017, one industry blogger was asking 'Will Volkswagen become #1 electric automaker within 5–10 years? It could.'

Müller was right. It is unstoppable. In August 2017, the cover of *The Economist* proclaimed the death of the internal combustion engine. Below the headline 'Roadkill', an illustration depicted an engine lying dead on the road, bleeding oil. Inside the leader proclaimed, 'The first death rattles of the internal combustion engine are already reverberating around the world – and many of the consequences will be welcome.' BMW and Jaguar Land Rover announced that all their models will have an electrified option by 2020. Volvo has gone one better, promising to electrify all its models by 2019, with the aim to sell only electric vehicles by 2025 – the same year that Norway, Austria and the Netherlands plan to achieve 100 per cent zero car emissions. India has plans to fully electrify all cars sold in the country by 2030. Bloomberg New Energy Finance predicts that the cost of ownership – including both the purchase price and the running costs – of electric cars will dip below ICE vehicles by

---

* Presumably aimed at the Yorkshire market.

2022 at the latest. China, the world's largest car market, with a unique influence over the global car industry, has signalled its intention to go fully electric too. Chinese firm Geely, which owns Volvo, was also responsible for the new plug-in hybrid London taxis: when in the London Ultra Low Emission Zone, the taxis will run on battery only, and then revert to petrol for longer journeys, or when the cabbies return home at the end of the day. The first cabbie to receive his keys for the new model on 1 January 2018 said it would save him £500 to £600 a month on fuel costs.

And then there's Tesla. The transition to zero emissions vehicles might be thanks, in part, to VW getting caught with its trousers down. But the popularity of electric cars owes as much if not more to Tesla. Before Tesla, electric cars were small, ugly but worthy – the yellow triangular doorstop types displayed on Brian's picture wall in Milton Keynes. But when the mercurial Elon Musk came along, a young entrepreneur and billionaire thanks to the success of his first company, PayPal, his aim was to produce an affordable, mass-market electric car. He told investors in 2008 that he was aiming for a price point of $20,000–$30,000 (£15,000–£23,000). But the chosen means of getting there looked nothing like a mass-produced family car. He started with the high-end sports car market. For the whole strategy to work, first the EV had to be made sexy.

In 2008, the company unveiled the Roadster, a sports car capable of accelerating from 0 to 60 mph in 3.7 seconds, and of travelling 244 miles in one charge.* Built by Lotus, its starting price of over $100,000 (£75,000) was hardly courting the mass market. The rave reviews it received on its proto-type launch in 2006 (such as the *Washington Post*'s paean, 'this is … more Ferrari than Prius – and more about testosterone

---

* Though arguably, since Musk launched his personal red Roadster into space with his SpaceX rocket in February 2018, both top speed and range have been wildly surpassed, with a top speed of 32.59 kilometres per second and an elliptical orbit around the Sun (aka range) of 250 million kilometres.

than granola') were the launchpad that the company – and all EVs – needed. Only a couple of thousand Roadsters were eventually made, but the statement of intent came across loud and clear. It paved the way for the Model S, the company's first electric saloon car, this time priced at the slightly more affordable $50,000 (£38,000). While looking less like a sports car, its stats were actually better than the Roadster's, with 0 to 60mph in 2.5 seconds and a range of 335 miles (540 km). It also had upgradable software. All Model S cars sold since October 2014 have Autopilot, which allows for limited hands-free driving thanks to sensors around the car that can detect road markings and other vehicles. Autopilot-enabled cars receive software updates as and when they are released, the same as software updates on your phone. Rather than trying to beat traditional cars at their own game, Musk told analysts in 2015, 'We really designed the Model S to be a very sophisticated computer on wheels.' It appealed to the tech enthusiasts, the speed junkies and the environmentalists – arguably three demographics that had never been combined before. By the end of 2017 it had sold over 200,000 worldwide. The Model S displaced Mercedes to become America's best-selling saloon car. In April 2017, Tesla had, astonishingly, surpassed General Motors as the most valuable US car company. Electric cars – or at least, Tesla cars – had become desirable, aspirational items. And in 2018, Musk's original dream began rolling off production lines: the affordable, mass-produced Model 3, with a starting price of $35,000 (£26,000).

Now Tesla is moving into the HGV market too, with the Tesla Semi, an articulated lorry with either a 300-mile or a 500-mile range, due to roll out in 2019. In Delhi, shortly after the prototype launch of the Semi, Shubhani, one of the leaders of the school mothers' campaign group, tells me: 'And of course *all* of us are in love with Elon Musk … Have you seen the huge trucks he made? I mean, *oh my God*!' The reaction from business was similarly gushing. Within a month of the launch, PepsiCo placed a pre-order for 100 Tesla Semis and food distribution company Sysco ordered 50.

Musk also believes that the lithium battery, used within all EVs, can solve the world's energy problems. Since 2014, the company has been building the biggest factory in the world (30 per cent complete by 2017) in the middle of the Nevada desert. The 'Gigafactory' produces lithium-ion batteries and, according to the Tesla website, its 'mission is to accelerate the world's transition to sustainable energy through increasingly affordable electric vehicles and energy products. To achieve its planned production rate of 500,000 cars per year by 2018, Tesla alone will require today's entire worldwide supply of lithium-ion batteries. The Tesla Gigafactory was born out of necessity and will supply enough batteries to support Tesla's projected vehicle demand.' Lithium-ion batteries don't transport well – shipping containers full of them rubbing together is not an enticing prospect, as that could cause them to explode. Any country going big on electric car batteries ideally needs to build their own. But for those countries who are not sure where to start, Tesla is again happy to step in and do it for them. Following frequent electricity blackouts in South Australia, Tesla tweeted a bet in early 2017 to the South Australian authorities saying that Tesla could build the world's largest lithium-ion battery for them, connect it to the region's 99-turbine wind farm, and solve its capacity storage issues for good – and if they couldn't build it within 100 days, they would do it for free. In an era when much political policy-making was seemingly bashed out in 140 characters or less, it was entirely fitting that South Australia took up the bet. The contract was signed on 29 September, and the world's biggest battery – now called Hornsdale Power Reserve – was completed on 1 December, with almost 40 days to spare (although in reality, Tesla had already set to work in July, but let's not split hairs). The 100MW/129MWh* battery bank, which takes up around

---

* The difference being power versus energy: so this battery array can deliver 100 megawatts (power) at any given moment, and can store 129 megawatt hours (energy).

10,000 square metres (12,000 square yards, or 2.5 acres) of land, can store enough energy to power 30,000 homes. Attached to the 325MW Hornsdale wind farm, it effectively means that wind energy can now be stored and utilised when the wind is no longer blowing, saving the state from further power shortages.[*]

But Musk's grandest vision is for every home to have solar panels and its own mini battery bank, which, combined with an electric car, forms a self-sufficient grid. When charged by rooftop solar panels – and Tesla also owns the solar roof tiles company SolarCity – the homes and cars become both the power station and the storage facility. There would no longer be any need for large coal or nuclear power stations at the other end of the power line. This is happening first in South Australia too, where the government is now building the 'world's largest virtual power plant' by installing solar panels and Tesla's 13.5kWh Powerwall 2[†] home batteries in at least 50,000 homes, beginning in 2019. Installation is planned over four years, and together the households will combine to create a virtual 250MW power plant. Energy from the solar panels will be stored in the Tesla batteries, and any excess energy will be fed back to the grid – and in the case of blackouts caused by the region's extreme weather, according to the South Australian government website, 'Powerwall can detect an outage, disconnect from the grid, and automatically restore power to your home in a fraction of a second.'

---

[*] Musk has even suggested – more theoretically than seriously – that to power the entire US you would just need 100 square miles of solar panels, 'a fairly small corner of Nevada or Texas', plus a bank of batteries '1 mile by 1 mile. One square mile. That's it.'

[†] Just to visualise how much energy that is, a standard rechargeable AA battery holds 2.4Wh. So a 13.5kWh Powerwall 2 is equivalent to 5,625 rechargeable AA batteries. It's also worth noting that the Hornsdale Power Reserve isn't one giant battery, but literally hundreds of Powerwall 2s connected together.

This is – if combined with electric vehicles – almost the perfect realisation of a clean air future. The Aussie BBQ will be the last smoke left in town. But Australia has two distinct advantages: an abundance of sunshine, and an abundance of lithium. The lithium-ion battery, invented in Oxford, may yet be our saviour, but whether we have enough precious metals on Earth to power the world with them remains an open question. The battery pack remains the inhibitor to growth, says Ian Constance, CEO, Advanced Propulsion Centre UK: 'The battery will be a massive differentiator … The rapid growth in the EV market means that there will be unprecedented demand for batteries, and as such vast production facilities are cropping up. Many more are going to be needed in the places where cars are assembled. Reports suggest that as many as 12 giga-factories are likely to be built in Europe alone by 2040.' The cost of the battery, he says, already represents about half the cost of an electric car.

The global scramble for lithium is on. Currently Australia produces most of the world's lithium, closely followed by Chile and China. The price of lithium carbonate rose from $4,000 (£3,000) a tonne in 2011 to more than $14,000 (£11,000) a tonne by 2017, raising some obvious concerns. Ally Lewis in York, for instance: 'The electric vehicle fleet will massively increase, but the unintended consequences could be who is making the batteries, and where and under what conditions, and what are they doing with all the waste? … You look at the volume of resources and the manufacturing scale that you need to come up with all these materials and it's quite staggering, it's not impossible, but it's a staggering amount of digging up that needs to be done. And we don't have a great track record of digging stuff up out of the ground without consequences.' Even more concerning, the world's supply of cobalt – another essential ingredient in a lithium-ion battery – comes almost exclusively from a single country, the Democratic Republic of Congo. But the industry view, according to Dick Elsy, CEO, High Value Manufacturing Catapult (a joint government- and business-funded research body) and former product director at Jaguar Land Rover, is more relaxed: 'Lithium is in fact in plentiful supply, fortunately, and it's not in very difficult parts of the world to get to. So, the battery industry accepts that lithium and copper and everything aren't in constraint, because the actual amount of materials used for batteries are surprisingly relatively small.' Elsy also believes that batteries will get incrementally smaller and lighter as efficiencies are found, pointing out that the Energy Innovation Centre at Warwick University has already made a battery with 70–80 per cent better energy density compared to those from the Tesla gigafactory. There is another major benefit, too: unlike liquid fuels that are pumped in and burned up, most of the lithium and precious metals within batteries can be continually recycled. How and how well this recycling is done will be critical. In February 2018, China's industry ministry issued new rules requiring carmakers to recover their vehicle batteries and set up recycling facilities and services. In India, one electric car entrepreneur, Manoj Kumar Upadhyay, is proposing an infrastructure of battery-swapping stations rather than battery-recharging stations. In his vision, when the charge is low you pull into a battery station, in the exact same way you currently pull into a petrol station, and swap the battery with a fully charged one. 'The target is to have many swapping stations and do quick battery swapping in under two minutes per swap,' Manoj tells me. 'In our opinion, this will be a faster way to increase adoption of EVs in India.'

It may also be possible that the lithium element of batteries could be replaced with sodium, one of the world's most abundant minerals.* A battery that uses a liquid sodium electrode was first proposed in 1968, but the membrane required to keep the electrodes sufficiently separated was too expensive. In the 2010s, several university teams around the world, including MIT, announced breakthroughs in the technology, and while none have yet reached production,

---

* Whereas sodium makes up 2.6 per cent of the Earth's crust, lithium accounts for only 0.007 per cent.

the theory now looks to be a practical reality. A prototype sodium battery has demonstrated an energy density of 650 watt-hours per kilogram – which would mean 650 miles of range for an electric car, twice that of the current best lithium-ion battery.

Electric buses will also have a huge impact on the air quality of a city, arguably much more so than private cars. Asian nations are leading this revolution, with almost 200,000 electric bus sales in China alone from 2016 to 2018. The city of Shenzhen has a fully electric bus fleet of 17,000 vehicles. The council in York (UK) has a fleet of electric single-decker buses already operating on the city's Park & Ride system and is trialling an electric double-decker bus on routes within the city – the first UK council to do so outside of London. Given the high volumes of NOx emitted by diesel buses, replacing a fleet with EV is one of the best things city councils can do to bring NOx levels down.

In business, electric fleets are beginning – slowly – to replace petrol and diesel for light delivery vehicles. UPS, one of the world's largest delivery firms, had more than 9,000 alternative fuel or electric vehicles by 2018, and pre-ordered 125 Tesla Semi trucks almost as soon as they were unveiled. Its president of engineering, Carlton Rose, announced that electric vans were now the same price to buy as diesel vans and were far cheaper to run. Many cities and countries are rapidly electrifying their train networks too. India, most ambitiously, is in the process of electrifying its entire 66,000km (40,000-mile) rail network by 2021; by the end of 2017, it was already halfway there.

The EVs on two wheels are arguably even more impor- tant for future clean air cities than the four-wheel kind, because you can fit far more of them onto a road. In London, Islington council are helping to increase the take-up of electric scooters for delivery businesses, including pizza delivery. During National Clean Air day, I visited a Greenwich council stand offering free rides on an electric bicycle – which, unlike an electric scooter, is simply a normal bicycle with a motor to supplement your pedal power. I give one a try, sceptically at first, as e-bikes have been touted as the 'next big thing' in sustainability circles for years, but have never really come of age. My suspicion is that cyclists like to pedal for fitness and non-cyclists would rather be in a fully motorised vehicle, leaving the e-bike isolated in a no man's land in between. But then I haven't ridden my own bike for ages, and as I whizz up a nearby hill with minimal effort, it's easy to contemplate how much more pleasurable future cycle rides would be if I had one of these. It even looks like a normal bike, so I exude a superhuman prowess as I nonchalantly pass a lycra-clad cyclist sweating every inch of the steep incline. When I cycle back by a different route, however, and find myself confronted by steps back to the Greenwich Centre, my superhuman façade is immediately shattered. The bike is bloody heavy. I can barely carry it.

But lagging at least 20 years behind the rest of the transportation field, once again, is shipping. At the turn of the twenty-first century, automobile fuel with 0.1 per cent or 1,000ppm sulphur was the norm in most developed countries (and is now as low as 0.001% or 10 ppm within the EU). When I talk to Simon Bennett, director of policy and external relations at the International Chamber of Shipping, he enthuses about the new regulations coming in for shipping in 2020: 'When we're using words like "game changer", we're not saying it lightly … in 2020 everywhere in the world shipping will have to use fuel with a content of 0.5 per cent sulphur fuel … At the moment [2017], you've got the situation where you could still burn the residual fuel with a fuel content in theory of round about 3 per cent sulphur in the middle of the ocean … you can continue burning residual fuel, the dirty stuff, provided you have an exhaust gas cleaning system, or what we call a scrubber.'* I ask if any major shipping

---

* Scrubbers are retro-fitted filter systems approved by the IMO and the EU and can cut sulphur dioxide emissions by 99 per cent. However, many of them are simply washed in the open sea, leading to water pollution.

companies are looking into electrification. 'Ships are just such enormous things so the concept of powering a ship by a battery, I don't know whether you've seen a merchant ship or just how big these things are, but … it's not quite the same as having a lithium battery to power a car.' I try a different tack – is there a demand from employees, from ship and port crews, to clean up the fumes they have to breathe in every day? 'That's not an issue. I don't think there's any particular health impacts from using one type of fuel over another, when we're talking about diesel fuel and the like. It's not an angle I've really thought about, in all honesty.'

There are some tiny green shoots, however. The Norwegian cruise company Hurtigruten launches the world's first hybrid battery-powered cruise ships, the MS *Roald Amundsen* and the MS *Fridtjof Nansen*, in 2019 – both 140m (460ft) in length, 21,000 tonnes, and carrying 530 passengers. Also in Norway (which incidentally has by far the highest rate of electric car ownership in the world) plans are afoot to fit a Bergen car ferry with a hydrogen-fuelled engine. The first fully electric cargo ship, meanwhile, has already set sail from Guangzhou, China, in 2017, travelling for 80km (50 miles) on just two hours of charge. Huang Jialin, chairman of Hangzhou Modern Ship Design & Research Co, the company behind the ship, told *China Daily*: 'The technology will soon be likely … used in passenger or engineering ships.' There was, however, a huge dose of irony in the ship's intended cargo. According to *Cleantechnica.com*, the all-electric cargo ship will primarily be used to transport coal to power stations along the Pearl River.

Arguably the biggest engineering challenge for electrification is in aviation. At LAX airport in California, Mary Nichols tells me of prototype battery-powered aircraft she has seen that can 'take cargo on short-hop trips … where you have small cities that are just far enough apart that they use airplanes rather than trucks to deliver mail and packages and so forth – they can actually operate on batteries that are recharged at one end of the trip.' There are various electric passenger plane projects too. Siemens is working on a hybrid electric passenger aeroplane, while the start-up Wright Electric – in collaboration with EasyJet – has designed a fully electric short-haul passenger plane, with the hope to launch by 2028. Uber are also working on UberAIR, a fleet of inner-city electric passenger drones, which it hopes to offer to customers in LA as soon as 2020. Several companies are seeking the same 'flying car' prize too. Chinese drone company Ehang released a video of its single-seater, 8-propeller electric drone in full flight in late 2017 (albeit without a passenger. Instead it delivered Christmas presents to a bemused-looking group of schoolkids in a park).

Electrification is happening, and it will replace fossil fuels. Even oil companies are starting to admit 'if you can't beat them, join them'. In October 2017, Shell acquired NewMotion, one of Europe's largest electric vehicle charging companies. NewMotion's website stated in early 2018: 'Our customers made a choice which now saves our planet more than 2,000 tonnes of carbon a year. It's time to go electric.' Coal India has side-stepped into large solar power projects. One of the world's biggest car parts manufacturers, Japan's NGK Spark Plug Co, has even announced it will soon no longer make spark plugs (which suggests a change of name might be in order), and will make EV batteries instead. Its senior general manager of engineering told Reuters that it was 'inevitable that the industry would at some point shift from the internal combustion engine to battery EVs'.

Something else that might be inevitable – but potentially less good news – is the rise of autonomous, driverless, vehicles. Just as most major car companies now have an electric car either in the showroom or in the pipeline, so most of them have a driverless vehicle project running behind the scenes. At the LCV car show, many were already looking beyond EV to driverless cars – or Connected and Autonomous Vehicles (CAV) as the industry prefers to call them. The 'connected' part refers to the ability of the cars to communicate with each other and with central control systems (for example, when temporary speed restrictions are put on a motorway, you would no longer need red signs flashing '40' above the road – they would simply communicate it to your car, which would automatically slow to 40). The Audi A8, on the roads from 2019, can drive itself at up to 37.3 mph, handling acceleration, steering and braking. But it still has a steering wheel and requires the driver to be alert. The cars being built by the likes of Google's Waymo, Uber and Baidu in China won't have a steering wheel, or even a driver's seat. Meanwhile, Tesla's CAV is arguably not so much being built as programmed. Elon Musk has stated that the same Model 3 electric cars that are already on the roads have sufficient hardware for fully autonomous travel. They will, in the near future, receive an optional upgrade to go fully driverless. Tesla is even working on an app-based system that will allow Tesla owners to send their cars out as driverless taxis when they are not using them, such as at night when they are asleep.

Modest little Milton Keynes is again miles ahead of the rest of the UK when it comes to autonomous cars. Inside one of the city's many anonymous grey tower blocks is the Transport Systems Catapult (TSC), a part-government, part-industry-funded R&D centre. It has licence to experiment with the kind of future-facing stuff that most government departments or businesses don't have the time to do, but are happy to fund. On the day I visit, Jaguar Land Rover have taken 'the pod' – an autonomous two-seater vehicle that moves at just over walking pace – out for a test run/walk. But I can do the next best thing – drive the pod simulator through a future Milton Keynes, weaving in and out of pedestrians. In the testing room, the pod simulator is connected to a moving platform and a VR headset. A bicycle has also been set up on rollers, to cycle through an imagined Milton Keynes full of CAVs, to see how cyclists might interact with them. In the middle of the room is a large octagonal space with a virtual reality (VR) headset and what looks worryingly like a black bullet-proof vest hanging from the ceiling, with gloves connected to wires. This space, I am told, is the 'Omnideck', a 6m (20ft) diameter, multi-directional treadmill that allows users to walk in any direction within VR scenarios. Martin Pett, head of visualisation, kindly talks me through it all, while his team members tap away on laptops on the periphery. 'There is a lot of engineering going on to make pods with millions of sensors on them, which will help the pods navigate,' he explains. 'But for me, I want to know what it's like for the end user, and that should then define how these vehicles behave and how they control them ... understanding how people feel about sharing road space with 400-kilogram vehicles that have their own mind and behavioural patterns.

To enter this virtual environment, I take the seat in the vehicle simulator and put on the VR headset. Suddenly I'm no longer in an office building but inside a two-seater car. There is a dashboard in front of me but no steering wheel. I appear to be in the same car park that I parked my own car in earlier that morning, only on a much sunnier day and with far fewer cars. Martin tells me to put my hands up in front of my face: they appear like stick man fingers, with white lines connected to red dots on the knuckles. I use my new white stick fingers to push a big green virtual button saying 'start' on the dashboard, and the pod moves forward. I go over a bump and I can actually feel it. I ask why there is no speedometer. 'Why would you need a speed reading as a passenger?' replies Martin. 'These questions only arise when you start to visualise these things.' He switches a setting and my hands change to ones with black gloves on, and then to human skin – pink arms slightly chubbier than mine are moving in front of my eyes, with the same movements as my own. Outside the pod, virtual pedestrians walk stiffly along the path. But despite the strange sterility of the scene, the movement feels real, and the sense of sitting in a moving vehicle without a driver or steering wheel is there. And, oddly, it doesn't feel odd. After the first few seconds it's just like sitting in the upper deck of a bus, or the front of a shuttle train at an airport; the sense of simply being moved to a destination is surprisingly familiar.

In the council, Brian Matthews believes that rather than driverless cars, the first CAVs in Milton Keynes will be like the one I experienced in the simulator: small pods on pavements, operating at slightly faster than walking pace:

'you have a quick interface with a pod and that pod delivers you to the door of where you want to be, so you don't need to park, you don't need to fight your way through the city centre. There's no written plans for this, but if we or another city wanted to stop cars going into the city centre because they're bad polluters or congestion is too bad … then the buses will sort of skim the city edge, meet a pod or meet a cycle hire or a very good walk route and get you to your last mile destination. That's what we are looking to put together.' With many large businesses headquartered in Milton Keynes, Brian believes take-up among commuters from the train station to their offices will be one of the first major uses of CAVs in cities like his.

After my VR experience I am snapped out of my trance and ushered into a more prosaic meeting room with Paul Bate, principal technologist and head of project delivery. 'When we talk about CAVs, Connected and Autonomous Vehicles, the "C" is very important,' he tells me. Paul believes that centralised urban traffic control centres could send signals out to the CAVs, and shepherd them according to their emissions credentials. He calls it a 'geo-fence', kind of like a congestion charge zone, but without the charge, able to flex in size according to the pollution conditions. 'This is a short-term solution while we still have fossil-fuelled vehicles on the road', he explains. But I ask Paul, could CAVs actually make congestion worse? If you can just sit on the back seat and flick through Facebook, do cars become more attractive than cleaner forms of transport, such as bicycles or electric buses? 'That is a potential outcome,' he admits. 'Again, it comes down to the "Connected" part – how do we manage those fleets, how do we optimise them to work to the benefit of the wider area rather than the one individual. Otherwise you are quite right, we could have gridlock with thousands of pods trying to get to the train station in the morning. If each vehicle works for one individual, you will certainly get that problem.'

Prototype CAVs shown at car shows have included sofa-style seating and TV screens. As tech website *theringer.com* puts it, cars are being 'reimagined as rolling living rooms'. Some have even suggested that CAVs will threaten the short-haul flight or sleeper train market, as you could lie back and sleep while you are carried from the door of your house to the door of your destination. A 2015 OECD (Organisation for Economic Co-operation and Development) study of Lisbon, Portugal, modelled future scenarios of the city based on the impact of shared CAVs called 'TaxiBots', compared to single-passenger, pod-style 'AutoVots'. In the ride-sharing TaxiBot scenario, supported by a well-running public transport system, 90 per cent of vehicles could be removed from the streets while still delivering nearly the same level of mobility. In an AutoVot scenario, where we don't share our rides but do grab some popcorn and watch our favourite films during our journey, the amount of kilometres travelled by car increases by 150.9 per cent. This may not necessarily mean more congestion – CAVs could in theory travel seamlessly bumper-to-bumper with no need for traffic lights – but it would mean far more road wear and tear, resuspended dust particles, and less road space for cleaner options such as bicycles and e-bikes. CAVs are also very greedy consumers of energy. The computing power required for all the sensors in an autonomous car has been likened to having 100 laptops plugged in and running simultaneously, and that's not including the actual propulsion of the vehicle. Ford's president of global markets has told investors that CAVs will be 'a significant drain on overall efficiency and fuel economy'. However, in Nissan's vision of the future, CAVs could drive themselves off to charging points at night, and then take themselves home again, topping up the home electricity at the same time. If this was responding to surges of renewable energy, then it could potentially more than make up for its additional computational demands.

CAVs have a lot of urgent questions that need answering, but, like it or not, they will be part of our electric future. The Waymo 'Robo-taxi', produced by Google, hopes to service paying customers in Phoenix, Arizona, in 2019, while Nissan is preparing its own version ready for visitors during the 2020

Tokyo Olympics Games. But both EVs and CAVs still have some competition. Liquid fuel isn't done yet; however, rather than being oil-based, it might be liquid hydrogen.

When I bought a nine-year-old Toyota Prius in 2016, it was pretty much the only game in town for an affordable second-hand hybrid. The Prius arguably kick-started the whole market. At the LCV Show, while impressed by Nissan's vision of an electric future, I was still rooting for Toyota when I got the chance to hear Tony Walker, Toyota's UK deputy managing director. Surely the company that first unlocked this market was way ahead of its competitors?

He started well. The three key environmental issues of our time, says Walker, are climate change, energy security and air quality: 'Toyota has sought to embrace these challenges by adopting hybrid technology and bringing it to the market over 20 years ago. This is in fact the 20th anniversary of the first sale of the Prius.' I know – big fan over here! I prepare to be amazed by what they're working on now, given a 20-year head start. 'Hybrids have the advantage of being very fuel efficient, and reducing CO_2 emissions, and at the same time emit very low levels of nitrogen oxides. And for Toyota, hybrid will remain our core technology.' In other words, remember that idea we had 20 years ago? Yeah, we're sticking with that one. I think of Kodak, who famously invented the digital camera only to keep it hidden for fear of damaging its camera-film business; the market changed so rapidly that by the time Kodak scrambled to get its own digital camera to market, its competitors had already usurped them. Toyota seems to be saying that we will always want a tank of liquid fuel inside our car, just like we wanted film inside a camera.

However, Toyota's vision for what that fuel consists of does have some strong clean air credentials. 'Toyota has been researching, developing and refining our hydrogen fuel-cell systems since 1992, resulting in the delivery of the Mirai – our first fuel-cell production model that was released in 2015,' Walker tells the conference. 'For the customer and wider society the benefits of hydrogen include … zero emissions, it only emits water at the tailpipe when driven. It offers a good driving range – around 300 miles on one refuel. And refuels in only three to five minutes, similar to a conventional vehicle.' Toyota only produced 2,000 such vehicles in 2016, and around 3,000 in 2017, but it aims to sell 30,000 a year globally by 2020. Given that the UK alone can expect around 2.5 million new car sales each year, this is hardly a game changer. 'So why does Toyota think hydrogen is so interesting?' says Walker, reading my mind. 'A very common question. It has the potential for zero carbon footprint if produced from renewable energy, and emits neither NOx nor particulates. It can be stored in large quantities, and for a long period, and can be transported over long distances. It has a high energy density – as everybody knows, higher than batteries – and as part of a transition to a hydrogen society, it can be used in other applications, for example in the home … for domestic heating, with adaptation.'

Part of me wants to jump up and shout, 'You are missing the transition – it is to an electric-based society!' Registrations of electric cars in the UK increased by 1,864 per cent between 2011 and late 2017, and every car company is now electrifying its range – thanks in no small part to the pioneering Prius. A third of car owners are actively considering an EV for their next purchase. Yet in early 2018 there were 82 hydrogen refuelling stations across the whole of Europe compared to nearly 140,000 public EV charging points (both stats according to the European Alternative Fuels Observatory). A 'hydrogen-based society'? Really?!

'A fundamental part of hydrogen's appeal is the fact that it can be produced from a variety of sources as well,' continues Walker. 'It can come from fossil fuel, it can come from biomass, it can be a by-product which is currently waste in some industries. It can come from renewable energy … With renewables such as solar and wind, you need a buffer to balance supply with demand, a role that we think that hydrogen can fulfil perfectly – it's a good means of storing the energy that has been generated through green and renewable sources.' Hydrogen is highly abundant – being two-thirds of water – but splitting water into hydrogen is known to be very energy intensive. The renewable energy argument here is that, rather than storing the surplus energy on windy or sunny days in batteries, use it to split water into hydrogen. Then we don't have to transition away from liquid fuel at all, but simply change the pumps at the petrol stations to hydrogen. Toyota is now working on fuel-cell buses, trucks and forklifts. 'We are convinced that both electricity and hydrogen will enable a more sustainable future society,' argues Walker.

According to Andy Eastlake at the Low Carbon Vehicle Partnership, whose duty is to remain 'technology agnostic' (as long as it's low carbon, he's a fan), 'Hydrogen has always been this "fuel of the future". It's zero tailpipe emissions, so that's great … you can produce hydrogen from zero carbon renewable energy – but actually you lose quite a lot of energy in producing the hydrogen, compressing it, storing it, putting it into the vehicle, converting it back into electricity. A hydrogen vehicle is an electric vehicle with a hydrogen fuel-cell range extender. So, it is good at long-range vehicles, trucks, things like that, where you need a lot of energy on board. But there is genuinely a discussion or debate to be had: do we put more batteries into a vehicle to have longer range, do we put hydrogen in a vehicle to have longer range, or do we just put more high-power infrastructure to charge the vehicles up more regularly? There are different ways of skinning a cat.' Hydrogen is essentially a hybrid car, then – the Prius, just with a different liquid fuel in it. 'You can put hydrogen into a combustion engine and burn it, but you are going to get NOx and all the other things,' says Eastlake. 'But a fuel-cell vehicle is an electric vehicle, you have a battery as well for your regen [regenerative] braking, and then you have a hydrogen fuel cell that produces electricity that feeds into the battery … You can have a diesel engine as a range extender. You can have a hydrogen fuel cell. You can have additional batteries. There may be other energy sources in the future. So, one of the messages I've been giving the government or anyone that will listen is let's not focus on the technologies – let's focus on the objective. And the objective is zero emissions.'

But if people think EVs have an infrastructure problem in the UK, with around 17,000 public charging points, then surely hydrogen, with only 12 fuelling stations, is a lame horse to back? 'It is a challenge, and that is a very valid argument,' says Eastlake. 'The hydrogen project – H2Mobility – have speculated that just 64 strategically placed hydrogen stations would allow you to create an infrastructure that could support the UK' – by which he means the UK's haulage sector, not private cars. 'One of the benefits of hydrogen is you can refuel very fast – you can get a lot of energy on board a bit like you can with diesel or petrol ... The other thing is the potential to repurpose the gas grid. The UK has a very good gas grid – long term we could look to repurposing that, and suddenly you have a distribution that is very, very effective for hydrogen.' The hydrogen molecule is smaller than methane, though, so can leak more easily; filling already ageing gas grids such as the one in the UK with leakier gas isn't entirely enticing.

California's plans to spend $2.5 billion to add 250,000 electric vehicle charging stations and 200 hydrogen fuelling stations by 2025, also suggests a strategy of EV for personal cars and hydrogen for commercial vehicles. When I ask Mary Nichols, she tells me, despite Toyota's protestations to the contrary, that 'There is a competition underway at the moment, I would say ... hydrogen today, we are only beginning to see what we can do with vehicles. There is one truck operating at the port of LA now, an enormous drainage truck that takes containers around on the port, and it's working quite well. But you know, it's one vehicle.' That said, she has herself recently switched to a Toyota Mirai hydrogen fuel-cell car, to discover what it is like for herself. 'It is very fun to drive, a really pleasant car, quiet and powerful,' she tells me. 'I fill it up at a couple of different stations which are reasonably close to my house, maybe a 10-minute drive ... It takes just about as much time as filling a tank in a normal gasoline station, and then I have about 250 to 300 miles on the tank. Unlike a battery electric car which really operates best in stop-and-start traffic, my hydrogen car, although it has regenerative brakes, it prefers a smoother, less stop-and-go situation, my fuel economy is probably not as good as it could be if I went to work by freeway every day ... The issue is really whether there is going to be enough gas stations to make it convenient for people to drive. We went from zero to 31 in California over the course of just a few years. It is already getting much easier.'

There are similar arguments for compressed natural gas (CNG) and liquid petroleum gas (LPG). As we've seen, CNG is a major fuel in India, but also in other parts of Asia, South America and the Middle East. LPG is available in over 1,400 filling stations in the UK alone. Both are relatively low-emission fuels compared to petrol or diesel. But both are by-products of the petrochemical industry and could never claim to be zero emissions. According to the UK Department for Transport's official advice, 'LPG vehicles tend to fall between petrol and diesel in CO_2 performance' while 'Local pollutant (CO, HC, NOx and particulate matter) emissions performance of well-engineered LPG and CNG vehicles is similar to that of a petrol vehicle.' They only offer slight improvements. While it makes sense for cities with both a big air-pollution problem and an existing CNG infrastructure, such as Delhi or Tehran, to increase the use of CNG in the short term, the long-term solution remains electrification. And, OK, maybe some hydrogen too. The same goes for bio-fuels. I could come in for some criticism for barely giving bio-fuels a mention in this book, but while they have their place – for example, running farm machinery using fuel made from the on-site crop waste – the place to burn bio-fuel (which is almost always bio-diesel) is not within cities. If we're serious about clearing the air, we have got to stop burning stuff within millions of engines on the streets where we work and live.

At the end of the LCV show, a bus ferries us back to the test track where a mile-long line of parked cars await, including mine, ready to be driven home by their solitary drivers. At the start of the mile, a single electric car-charging station stands empty, waiting hopefully for an occupant. The charging point is hooked up to a diesel generator. Plumes of PM and NOx-filled heat shimmer above it. The image perfectly, depressingly encapsulates the dangers that lie ahead in the fight to clear the air. If we get it wrong, we end up travelling in separate driverless cars powered by diesel generator sets, while electric ships ferry the coal to the power stations.

CHAPTER NINE

Road Rage

Scandinavia always makes me wistful. The Finns are quick to point out that they are Nordic, not Scandinavian. But the same beautiful melancholia unites the whole peninsula. At Helsinki airport, despite the late Autumn warmth, the bus transfer proudly displays pictures of Finnish ski-jumping champions. I board my quiet electric train and am soon gliding through tall, pristine pine forest. This gives way to city buildings, malls and university campuses, a blur of whites and greys, like a rapid flick through a 1990s Ikea catalogue. I take my Egg out of my cabin bag and turn it on. When I had last used it on my early-morning train to the airport in England, it had registered in the 30s for PM2.5 $\mu g/m^3$. Now it is $1\mu g/m^3$. At one point, in a tunnel, it excitedly climbs to $7\mu g/m^3$, before getting giddy and falling back to $5\mu g/m^3$.

I've come to Helsinki because it has a strong claim to be the cleanest-air capital city in the world. But also, amazingly, it isn't resting on its laurels – in fact it is doing more than most cities to clean up what little air pollution it has left. And number one on its 'to do' list is to make the private car a relic of the past.

I message Sonja Heikkilä when I'm on the train. Sonja found herself the accidental poster girl of the clean mobility movement when her college thesis went viral – at least among international transport community types – in 2014. It coined a new term, 'Mobility as a Service', or 'MaaS', as a concept of ownership-free, multi-vehicle transport that could replace the dominance of the private car; today there are entire international conferences dedicated to 'MaaS', all thanks to Sonja's college thesis. The central premise is this: why own a car, or a moped, or a bicycle, or even buy a train ticket, if you could instead access whatever you need via a single swipe card or app? Why attempt to make an entire journey from A to B using just one form of transport, and then have to return from B to A using exactly the same one? Why needlessly pollute, just because you own something?

Outside Helsinki train station, electric trams mix with cars and bicycles. It looks busy, so I get my Egg out again: astonishingly, it's still only $1\mu g/m^3$. Sonja arrives at the pick-up point in front of the station tentatively driving a pristine white electric BMW i3. The company she now works for, the insurance firm OP, has embraced the MaaS idea and brought the car club DriveNow to Helsinki. Car club members have access to a fleet of cars dotted around the city, rather than needing to own one. Sonja is keen to meet me in one of its flagship EVs, though she admits she hasn't driven it many times before. 'Is it on?' she asks, as she tries to reverse. 'It's hard to tell. No, it isn't on.' She presses another button and there is a beep in lieu of an engine sound. 'The i3 is probably the most popular car in the external fleet,' she tells me as we set off. 'I can never find an i3.' There is a simple blue line on the dashboard display showing it has 160km of charge left. 'I think it's quite limiting if you need to either choose to live without a car and then are restricted to where you can go, or you choose to buy a car but are restricted to those costs that come with driving and car ownership,' she tells me, as we drive quietly through Finland's quiet capital. 'Even when you're not driving it you need to pay for the insurance, everything. You are not free ... I wanted to create this concept where you don't need to be dependent on anything, you can choose to use what you want at that moment ... I am not dependent on the car or bike that I took with me from home or work – I can change my plans during the day.'

OK, so 'MaaS' sounds like – and is – corporate jargon. But its central premise is one that all cities with clean air aspirations will need to embrace: we need to quit our car habit. This is part of the clean air blueprint. Electric cars are great, but fewer cars are even better. Remove the traffic, and the pollution goes too. In Newcastle, UK, when city centre roads were closed to traffic for the HSBC UK City Ride bicycle race in 2016, levels of NOx fell by 75 per cent.

Sonja's thesis adviser in 2014, Sampo Hietanen, has since set up his own company, MaaS Global. Its 'Whim' app allows subscribers to use many different forms of transport within Helsinki (and increasingly other cities, including Birmingham, UK) simply via an app. I decide to use Whim during my stay in the city, to experience it in practice. When Sonja drops me off at a Metro station, I'm a bundle of bags, papers and Dictaphone, with no idea where I actually am, so I look to Whim to tell me what to do. Top of its list of suggestions is a five-minute walk around the block to catch a tram. As I am standing directly outside the Metro stop, I don't want to do that, so I scroll down the list to find the Metro option and touch 'start journey'. There is no ticket, only the promise that the 'ticket will be automatically generated at the start of the journey'. When I get onto the train I'm slightly worried that it hasn't worked. Before I even have time to check, I'm met by a ticket inspector. 'Er, I'm using Whim? I say apologetically, and proffer my phone screen, hoping he'll know what to do. 'Whim?' he replies, expressionless, as if he'd never heard of it. This is not a good start. We both stare blankly at my phone. I weakly prod at the ticket button again. This time a green symbol appears saying 'HSL Seutu, valid for 60 min'. 'Yes, OK,' says the inspector, and moves on. I relax. I also have no idea how much that just cost me. But compared to struggling with a ticket machine, it was quick and easy.

I emerge in the city centre by the waterside. My Egg now wavers between 12 and 19µg/m^3, but never once breaks the WHO's 20µg/m^3 health advisory limit (except if I pass someone smoking, in which cases it can easily shoot up to triple figures). Walking down Kalevankatu Street, the six-storey brick buildings and cafés remind me of Manhattan. It then opens out into a boulevard that is more like Paris, except for the trams going up and down. There are cars too, but they are outnumbered, in passenger terms, by bikes and buses. The Egg reads just 9µg/m^3. Then 8µg/m^3.

MaaS Global's new penthouse office is literally being unwrapped when I arrive. The metal buzzer still has protective plastic that's yet to be peeled off and the elevator is internally lined with cardboard, like a parcel waiting to be opened. When I arrive the CEO, Sampo Hietanen, is regaling some staff members with a story from a recent trip to Japan – he seamlessly changes into English to include me. 'The toilet', he says, 'had a washbasin above it – you wash your hands, and then that "grey water" is used to flush the loo. Perfect closed-loop sustainability!' He offers me a coffee and ushers me in to a glass-walled meeting room. 'There is a problem of more and more dense cities. The normal curve has been as GDP rises, the ratio of people owning cars rises as well. Cities are jammed, they cannot move,' says Sampo. He, by contrast, is so full of energy and enthusiasm that he continues to move even when sitting down. 'So, we have two ways of trying to solve this issue: one is looking at the map and start banning things. But in a free democratic world, and even in other places, that's pretty hard – I'm not a big believer in what Oslo or Barcelona are trying to do. It is easy to say politically "by 2030 we are going to ban cars" – they are not going to do that. Unless – and this is the other solution – unless there is something better out there. If you want to focus on the issue of emissions ... you need to look at the individuals and ask, why did they buy that car ... if the paradigm of how we plan our mobility changes from owning a vehicle to owning our own operator or service provider, that changes the whole system.'

Car ownership and driving licence applications are already going down among young people across Europe and the US. Sampo believes this is not just because insurance costs are going up, but that ownership is 'seen as a drag'. In mid-2017 the regional public transit agency for Austin, Texas, piloted the Pickup app, allowing users to request a ride via public transport to anywhere within its service zone from their phones – in effect, an on-demand bus service. Similar services have started popping up around the world, such as the

Citymapper Smartbus in London. In Milton Keynes, Brian Matthews is already thinking about a driverless minibus system that gets you to where you want in a very similar time to a car, and the same quality that you experience in a car, so you get your personal space, so you're not crammed into a very small space ... it's the experience that will make it work and challenge the car.'

Another alternative is to get people to share their cars. This is an essential part of the MaaS idea, such as Sonja's latest car club scheme in Helsinki. At the London Assembly, Leonie Cooper is a big fan: 'It's about trying to get people to switch to electric car clubs, and sharing their cars and being more efficient with car use ... I think joining a car club is the way forward. Although they haven't got their own car, they have access to a car – so on that occasion that they need to take rubbish to the dump, or go and visit their aunt in Suffolk ... if they are a member of a car club, for some people it would be a lot cheaper. Owning a car is not cheap. I think we need to start presenting people with the evidence about the calculation between if you don't have a car – not paying road tax, no insurance ... And if you join a club with access to all kinds of vehicles, not just one.'

At the LCV Show, car-sharing was high on the car industry's agenda too. According to Konstanze Scharring, director of policy, Society of Motor Manufacturers and Traders (SMMT), 'exponential growth' of car-sharing is 'expected in a very short time-frame ... We expect membership of shared services to increase [in the UK] by 2.3 million by 2025.' Car-sharing operators – including Zipcar, DriveNow and Enterprise CarShare – currently have more than 200,000 members across the UK. If the 2.3 million forecast is correct this would remove an estimated 160,000 private cars from UK roads over the next eight to ten years. According to industry analysts Frost & Sullivan, app-based taxi services such as Uber and Lyft could potentially remove about 10.41 million vehicles from the road worldwide by

2025. This, said Scharring, heralds 'a significant change in the future of mobility'.

Whilst at LCV, I ask Andy Eastlake about the elephant in the room. If the end goal of car-sharing is fewer emissions and fewer cars on the road, that is surely the opposite of what every car company at this show actually wants? 'That is one of the end goals, I would absolutely agree,' he says. 'We have a number of key evolutions that we've got to go through. So, we've got to get the cars to be as efficient as possible, to use as little energy as possible. Whatever engine they are using has got to be zero carbon and zero emissions at the tailpipe – at the moment the clear leader is electric vehicles. So OK, we've got zero emissions mobility. What we aren't solving with that is congestion and space. If by 2050, rather than having 35 million petrol and diesel cars on the road we have 40 million electric vehicles on the road, we will have failed … we will be gridlocked. So, what we've got to do is change the whole model of mobility and ownership. Rather than having a vehicle that costs you £25,000 and sits static and idle 95 per cent of the time, maybe we want a vehicle that costs £35,000, but is sweated as an asset and used 65 per cent of the time. We could deliver the same number of miles with half the vehicles.' But the OEMs (Original Equipment Manufacturers) want to sell *more* cars, surely? 'You're selling better cars, at a slightly higher price, and the services that go with them – the cost of the mobility,' says Eastlake. 'And given that they are used more frequently, they will be replaced more frequently too.'

To reduce emissions in our cities, the private car in the driveway can no longer be the automatic go-to option. People often use the car because it is there, it is easy and they are already paying for it. Car clubs put cars on an equal footing to other mobility options, and ask the question: what is your best way of getting from A to B today? They are also one of the quickest ways of electrifying road transport. Not everyone can afford an electric car, but everyone who can afford to buy an old car, can buy a car club membership instead. The Paris AutoLib was the biggest electric car club in the world before its plug was pulled in 2018, but its spirit lives on in the many cities around the world that have since copied it[*]. Wrocław, Poland, has 200 Nissan Leaf electric cars available through its Vozilla car-sharing club. The 'e-share mobi' in Japan is available on a membership-free, pay-per-use basis – the cars are also available with the new ProPILOT autonomous technology, making this a clever way to familiarise people with both EV and autonomous driving. The original company behind Autolib in Paris, Bolloré Group, has also gone on to run similar schemes around the world including Indianapolis (BlueIndy), Los Angeles (BlueLA), London (Bluecity) and Singapore (BlueSG).

But while electric car clubs are part of the clean air blueprint, weaning people off cars altogether is even better. This is not about removing all cars from the streets, but simply reducing the number of them, cutting into their dominance. Mark Watt, the executive director of C40 Cities Climate Leadership Group, wrote in his blog in 2017, 'Ultimately, private cars will never be the best climate and clean air solution … While electrifying our vehicles is an important step in tackling air pollution and climate change, citizens will ultimately need to move beyond private cars and shift to mass transit – buses, trains, car share – and good old-fashioned walking and cycling … making our streets safer, quieter and more pleasant places to be.'

'What will the MaaS city of the future look like?' Sampo asks himself, as he talks, ostensibly to me. 'The tube [London underground] is the backbone of London, for example. But then spanning out of that you would have hubs of multiple services – from the tube station you plug in your Segway share or your car or ride hailing [such as Uber] or CAV or

---

[*] Not least in Paris itself. In summer 2018, its electric car share market was opened up to multiple competitors, with Renault launching its own 'dockless' scheme, Moov'in, in September 2018. The fleet of Cityscoot e-scooters also grew to 6,000 by 2018 and 10,000 by 2019 – far bigger than Autolib ever was.

your drone, which connects you to other hubs. What do you currently see when you come out of the tube? A roundabout. So, if we want a new kind of future you have to build the infrastructure.'

There's a line in a WHO report that nicely sums it up, saying, 'Poor urban planning, which leads to sprawl and overdependence on private vehicle transport, is a major factor in urban emissions.' Sampo too, who started his career in urban planning, says, 'The sooner we start building on that future the sooner it will come.' Some cities are starting to do just that. Madrid's mayor, Manuela Carmena, plans to kick private cars out of her city centre. On Spanish radio she announced that the city's main avenue, the Gran Vía, will only allow access to bikes, buses and taxis by the time she leaves office in May 2019. This was in addition to a previous announcement that all diesel cars will be barred from entering Madrid by 2025, while 24 of the city's busiest streets are already being redesigned for walking rather than driving. Not wanting to be outdone by its regional rival, in September 2017, Barcelona created its first car-free 'superblock', covering 15,000 square metres (160,000 square feet) of Barcelona's El Poblenou neighbourhood, with six more superblocks, each housing around 5,000 to 6,000 people, to follow in 2018: the plan is estimated to lower pollution levels nearly threefold, simply by giving pedestrians and cyclists priority over cars. Roadside parking within the superblocks will go, making space for street games, sport and even an outdoor cinema.

Professor Chris Griffiths at St Barts Hospital is co-founder of the campaign group Doctors Against Diesel. He tells me, 'It's important to remove diesel from the road, but … it's not just about diesel, it's about reducing traffic, making city centres an environment where kids can walk to school or cycle to school instead of going in cars, they can play more, get more exercise, be less obese. Living in a city can be much more pleasant than it is now … no diesel, no internal combustion engines, less traffic, better transport systems. That means that the quality of life of people in cities is much improved.' In south London, Greenwich Councillor Dan Thorpe gives me a great example of this, called the 'walking school bus'. Just like a traditional school bus, it stops at regular points along the route to school, picking up kids at each stop – but unlike a traditional bus, there is no vehicle, only children in high-vis jackets walking hand in hand, with two designated adults – typically teachers and/or pre-allocated parents – at either end. 'Some schools have them on a daily basis,' he tells me. 'They have posters and slogans to spread the message. In the winter it gets a bit more tricky. But that kind of mass action is one way to make progress … We have an annual festival where we close off the streets for two days, it's amazing the sense of freedom it gives everyone.'

While public transport is crucial – and far better than individual electric cars – it is the transport that we can do with our own two feet (also known as 'active transport') which makes for the cleanest air. According to research by Delft University of Technology, a 3.5-metre-wide (11ft) road lane can transport only 2,000 people an hour in cars, compared to 14,000 cyclists or 19,000 pedestrians, before you even take into account the space saved on parking.

But the battle for that road space is very real, and very bloody. As the number of cyclists increases in major cities such as London and New York, so does the resentment between road users. Lives and limbs have been lost. In New York City in 2011 there were 754 crashes between motor vehicles and bicycles, killing three cyclists and injuring 755 people (of whom just 10 were vehicle occupants). By 2016, the numbers had greatly increased: 4,592 people were injured in 4,574 collisions with motor vehicles, in which 18 cyclists were killed (no motor vehicle occupants were killed). Most are genuine accidents, but not all. Headlines such as 'Pro cyclist knocked out by punch from road rage driver' (28 March 2018, *road.cc*), and black cab drivers 'wage all-out war on cycling' (22 March 2018, *London Evening Standard*), show the genuine hatred felt. While less shocking, a post on Twitter in April 2018 by the artist – and cyclist – Grayson Perry sums up the everyday, brewing antagonism: 'If you are the woman driving the black Citroen C3 who rammed me onto the pavement on Barnsbury Road this morning, fuck you.' There are Twitter accounts dedicated to posting such interactions daily, from both sides of the divide.

During Paris's car-free day, the same simmering resentment threatened to boil over. Traffic was only down by 52 per cent that day, with many cars and vans simply flouting the 'ban'. Police and volunteer organisers stopped some, but they were overwhelmed. The exemption for taxis and Uber drivers – which can easily look like private cars – also made it hard to distinguish genuine exemptions from laissez-faire locals. Mariella Eripret, one of the community organisers, told me that, 'Some of them, especially taxi drivers, were very aggressive ... It often happens like this ... they are angry when hundreds of cyclists take a big part of the road, which forces them to drive slower, or to wait longer at the traffic lights or the crossroads ... other drivers drove very fast, taking advantage of having less traffic ... There were still too many cars, and drivers did not respect the limit of 30 kilometres per hour.' Speaking to the *Guardian*, Christophe Najdovski, the deputy mayor in charge of transport, admitted: 'We have to change people's attitudes and behaviour.'

But there is an answer to all this, and one hinted at in earlier chapters: separate the cyclists from the vehicles, by building segregated bike lanes, and close off more roads to cars permanently. London has an average of 1.1 deaths per 10,000 cyclist commuters, which is better than New York's 3.8.[*] But this isn't about the number of cyclists each city has. Copenhagen and Amsterdam, two of the highest cycling-density cities in the world with around half of all commutes taken by bike, have just 0.3 and 0.4 deaths per 10,000 cyclist commuters, respectively. Both London and New York have a higher percentage of cycle helmet-wearers, too. The

---

[*] Which is scarily similar to the death rate among serving UK army personnel in 2014 of 4.3 deaths per 10,000 soldiers.

difference is that bikes are given their own dedicated lanes and roads away from trucks and cars. In Copenhagen, Denmark, over 50 per cent of all trips within the city are now taken by bicycle and 30 per cent by public transport, with just 25 per cent taken by car.

Helsinki is now trying to copy Copenhagen's success. Using the Whim app I step onto a bus that takes me to the Environment Centre, the Helsinki council department responsible for parks, recycling and air quality. It's on the outskirts of town and I have barely left enough time to make it for 5.30 p.m. as arranged. Everyone flashes a card as they board, except me clutching my phone. It strikes me how much quicker a cashless journey is than the not-so-very-old days of everyone tipping coins out of their pockets. Soon, there won't even be a card to flash. The bus departs and arrives exactly as Whim says it would, at 5.38 p.m. I apologise to Esa Nikunen, director general of the Environment Centre, and apologise again when I realise he is the only person left in the building. Finnish working hours typically end at 4 p.m. 'In Helsinki we are promoting the use of bicycles and walking and public transport quite a lot,' he tells me. 'The traffic planning is taking this quite seriously, they are building quite a lot of new bicycle lanes every year to Helsinki city, they are closing some streets from private cars in the city centre.' He tells me that the main road I took the bus on will soon be closed entirely to private cars, although the buses will still run. 'Many people are not very happy about it,' he says. A new €150 million (£130 million) bridge connecting the centre to one of Helsinki's many island suburbs is also to be built, 'and it's not going to be open to private cars at all, so it's just for public transport ... tram, bicycles and walking ... you have to drive about ten kilometres but the bridge is less than two kilometres long.' So, is the idea is to make non-car transportation the easier, more attractive option? 'Yes, that's right, that's the main tactic ... they're building more houses there, and then of course it's more attractive if you have a fast connection to the city centre from that new area.' I later learn that three such bridges are planned – known as the three

Crown Bridges – connecting to residential islands, all of which will be free of private cars.[*]

'The difference between Copenhagen and ... others is the service promise from an infrastructure point of view,' Sampo also tells me. 'When I biked in Copenhagen the best part of it was that when I turned left, I knew that the bike lane would still continue. Anywhere else it can stop any time and no one cares. It would never happen to a car.' Sami Sahala, chief innovation advisor at Helsinki city council, tells me that Helsinki has only recently embraced cycling. The city bike hire scheme has only been running a couple of years. 'But it's been hugely popular. This summer even though the weather wasn't that good, on the best days every 1,400 bicycles (there was 1,500 but we gave 100 to the neighbouring city) were used on average for 11 trips per day. That really is at the top end for cycling schemes. So now we seem to have become a cycling city as well ... at the same time we have been building cycling highways called "Baana". There used to be a rail line going through the city centre that has been disused for many years, and then somebody had the idea about eight years ago to turn that into a bicycle lane, and that was a big hit. There weren't huge numbers at first, but in time people starting realising the amount of time they could save when they needed to cross the city centre.'

I spend that afternoon trying out the Helsinki city bikes. Keen to get onto the nearest Baana, I pre-register online (it wasn't yet available on Whim) and walk to find the nearest bike docking station. The official map I download of Baana routes from hel.fi however is simply a pdf, and it's hard to work out exactly where I am, so I decide to just ride on the nearest path and see where it takes me. I quickly reach an

---

[*] This isn't only happening in Scandinavia (sorry, 'the Nordic countries'). The largest car-free bridge in the United States, Tilikum Crossing Bridge in Portland, OR opened in September 2015, with dedicated light rail, streetcar (tram) and bus lanes, plus paths for cyclists and pedestrians.

industrial area, riding by a busy three-lane carriageway of cars. This isn't quite the lush, green picture of Helsinki I had in mind. But soon I am on a bridge, cycling towards a marina full of little boats moored on the right bank; on the horizon, a yellow glow as the sun breaks through the clouds out at sea, and an ancient-looking sailing boat is silhouetted. As I cycle along a now blissfully easy, segregated bicycle lane, I almost want to wave and shout 'isn't this great?' as I pass other cyclists. Unlike in the UK, cyclists and pedestrians and cars seem to co-exist without animosity. Each accepts the other. When I cycle back towards the central station, I discover the original Baana: the old railway track cutting through the city centre, under numerous bridges, now a two-way cycle route and footpath. There are even digital information signs, similar to motorways, although I have no idea what they are telling me. Up on either side, above the railway-cut cliffs, the city roads are stuffed with slow-moving cars.

There is a definite 'if you build it they will come' aspect to cycling infrastructure. In the Netherlands, an entire country criss-crossed by segregated cycle lanes, 50 per cent of children cycle to school every day; in Greater Manchester, UK, it is less than 2 per cent. When London began building its cycle-superhighway scheme in 2010, it was pilloried for just being 'blue paint', easily flouted by drivers, and disliked by cyclists due to the lack of protection (i.e. they continued to die). In 2015, however, based an International Cycling Infrastructure Best Practice Study, the approach changed and London began physically separating the 'blue' cycle lanes from cars with concrete kerbs, narrowing the road lanes in order to make space. Right-wing newspapers published photos of empty cycle lanes next to gridlocked traffic, such as the *Daily Mail* front page headline on 5 October 2016: 'Cycle lane lunacy! The new blight paralysing Britain' (yes, that was a real national front page). But people previously too scared to cycle in London – due to the very real risk of death – could now cycle in confidence, and the lanes soon filled up. By early 2018, the *Mail*'s photographers would have to be up very early indeed to find an empty cycle superhighway: in the morning rush hour, between 7 and 10 a.m., cyclists account for 70 per cent of all traffic going over Blackfriars Bridge into central London.

Laura Parry at Islington Council is trying to get commercial businesses out of vans and onto bikes, too: specifically 'cargo bikes', with fitted boxes on the front to transport goods. It sounds lovely in theory, but are businesses really tempted? 'We've so far converted 13 businesses onto cargo bikes,' she tells me, only weeks into the trial. 'The way a lot of businesses are currently doing deliveries just doesn't make sense – I met a wholesale and retail business recently in Islington who had one shop near Angel and one near Oxford Street [around 2.5 miles apart]. Whenever they needed to deliver to a customer they would have to get their Oxford Street van to drive up to Angel and then deliver to customers just two streets away. So, we are helping them to switch to a cargo bike now because it was just not a good use of their time, resources, or money … they have five members of staff and they are all keen to take turns on the cargo bike. They are really excited. It is good for their customer relationships when they turn up on a bike … it's definitely good for advertising.' The food delivery company Deliveroo, for example, uses predominantly cyclists rather than motorbikes. Deliveroo works with over 15,000 self-employed bicycle delivery riders across the UK and 30,000 globally.

Given the dangers of exposure to road pollution, though, is it fair to encourage more people to cycle or walk on the roads? A large multi-country study led by the University of Cambridge in 2016 found that at PM2.5 levels of $22\mu g/m^3$, the benefits of physical activity 'by far outweigh risks from air pollution even under the most extreme levels of active travel'. In 99 per cent of polluted cities the health benefits of cycling outweigh the negative effects, said the study. Even at PM2.5 levels above $100\mu g/m^3$, which would be considered high almost anywhere (except Delhi), 'harms would exceed benefits after 1 hour 30 minutes of cycling per day or more than 10 hours of walking per day'. A Danish study in 2001 followed drivers and cyclists on the same routes and found that the drivers were exposed to 2–4 times the amount of PM and

BTEX (benzene, toluene, ethylbenzene and xylene) compared to the cyclists, who benefited from open airflow rather than stewing in stagnant air inside a car. It also found that children travelling on the back of a bicycle inhale significantly lower concentrations of pollution than those in the backseat of a car. A study in Montreal found that separated cycling lanes had a significant impact on personal exposures with a 12 per cent decrease observed for black carbon, compared to cycling within traffic. Cyclists and joggers are typically fighting the same air pollution from a much higher base line of general fitness; inactive people are asking their bodies to withstand those pollution levels with immune systems far less equipped to do so. The conclusion of the Cambridge study was that the benefits of walking and cycling, even in polluted cities, by far outweigh risks from air pollution, even under the most extreme levels of active travel: where walking or cycling replace a car journey, the benefits increase even more.

While in Delhi, I asked Dr Sharma what might be a sacrilegious question within the walls of the Central Road Research Institute (CRRI) – is part of the solution to the city's pollution problem to have fewer roads, and more cycle lanes and pavements? 'If I understand you properly you are talking about the non-motorised transport, pedestrians and cycling? In Delhi it is quite difficult because of the lack of facilities … it's too dangerous.' Delhi Metro stations disgorge a flood of passengers out onto broken pavements and busy highways, meaning they still need to catch a taxi to get where they need to go. 'You call it last mile connectivity. *Teek hey.* Theoretically it is a good idea. But it means infrastructure, infrastructure means investment, and Delhi Metro last mile connectivity has to be made … but the issue is always land acquisition. Everybody whose land is being taken and people are compensated. In India, people will die for their land. If it is compensation, they want hugely disproportionate compensation, they will go to court – I tell you we are too democratic – and in court it will take one or two years.' The Delhi Metro is 'the pride of India', says Dr Sharma, and without it pollution would certainly have been worse. But public transport here – as in many parts of the US, too – has an image problem. When I ask Dr Sharma if he takes public transport himself, he laughs. 'You will be surprised – I have done at least 20 projects for the Delhi Metro related to the environment as a consultant, as an advisor. But in my whole life I have not travelled by Metro more than two or three times.' I tell him I'm very surprised. 'Yes. The reason is last mile connectivity. Second, it's too overcrowded.' He prefers 'the comfort level' of his car, he says.

When I meet the Delhi-based author Rana Dasgupta later that week, he tells me, 'the use of public transport is directly connected to means. If you have the money to run cars for all the members of your household, there is no way your household will travel on Metro or buses – least of all buses.' He tells me about an attempt during the Delhi Commonwealth Games (2010) to re-engineer some of the major roads with bus-only lanes and dedicated cycle lanes. 'Of course this city is obsessed with position and hierarchy, the idea that poor people could be occupying these car-free areas of the road while everyone was queuing up, it didn't appeal at all,' he recalls. 'They all started driving in the bus lanes, and the buses got caught up in the traffic. So the city authorities employed people full-time to stand at the road divide, trying to divide the traffic, holding "buses only" signs up to the cars. Meanwhile the cycle lanes were separated from the roads by flimsy bollards – those bollards were knocked down in no time, and the cars started driving down the cycle lanes, which is what you see now … they have just completely given up on the idea.' The car, he says, remains fundamental to Delhi, and how Delhi-ites see themselves. 'You quite often see marriage ads [in the newspapers] where it will say "Groom, 25, drives Honda City XLS …" because it is an extremely good gauge of where somebody sits in the social hierarchy … It also signals your power on the road … people shoot each other, literally, over car parking spaces. There is no neighbourhood fight so big as over car parking space.' As he speaks there is a loud parping of a car horn outside.

Walking and cycling are actually very common means of transport in Delhi, 'but it is confined to the poorer 50 per cent of the city', says Rana. 'I had a friend who decided that she was going to sell her car and cycle everywhere ... But people actually stop and stare at wealthy people on bikes, because it is such an unusual sight. They are used to bikes designating a certain class.' Vandana, my B&B host, strongly advises me not to 'get on anything with two wheels. It's like a death sentence. I have so many foreign guests, young people ... telling me "we are going to cycle" – but really? You will be killed.' Yet she recalls a happier time when 'we all cycled when we were kids, we all had bikes ... I don't even cycle from my house to the Defence Colony market now [less than 1km]. I sold my bike.' Shubhani Talwar, a clean air campaigner, also laments the loss of cycling: 'What we need back for our city is that culture of cycling.' But, she says, it is a 'chicken or the egg' situation: 'Should we cycle when there is no cycling track?'

Carla Stephan, a medical doctor from Lebanon, now works in London for the public health charity, Medact. She tells me that Lebanon's pollution is similarly caused by having 'a lot of cars. For example, my family is five people and each of us has a car, so we have five cars – this is normal within Lebanon. Even poorer households will have a car for every person that's working, because there's no other way that you can go to work. There's no real public transport ... We used to have trains and we used to have buses, but the car lobby won over many developing countries ... we have so much congestion.' But some developing and middle-income countries have cracked the code. Bus Rapid Transit, for example, is distinct from normal bus systems in that they have their own dedicated roadways which no other vehicles are allowed on. It works like a metro system but on the surface, or a tram system without the tracks, and is therefore a far cheaper alternative to both. It was pioneered in Curitiba, Brazil, in 1974, and soon spread around the world including in Colombia, Turkey and Iran. Mexico City's Bus Rapid Transit system Metrobús, launched in 2005, is the longest such system in Latin America, serving 800,000 passengers per day and around 180 million passengers annually. But crucially it replaces the need for a car for many people. It has been estimated to have caused a 15 per cent shift from cars to public transport, or 122,000 fewer daily trips in private vehicles. Metrobús alone has been estimated to reduce annual NOx in the city by 690 tonnes.

As well as showing the world how to run Bus Rapid Transit, Mexico City has also led the 'greening cities' movement. This comes next on my clean air city blueprint: pollution-busting cities need lots of green space and vegetation. Trees, plants and grasses eat up carbon dioxide and filter out particulate matter. In Mexico City in 1986 there were only two square metres (2.4 square yards) of green space per inhabitant. By 2016, there were over 16 square metres (20 square yards) per inhabitant.[*] As well as increasing park space and tree planting, as of 2015 the city had over 35,000 square metres (42,000 square yards) of green roofs (rooftops planted with a dense covering of low-maintenance, hardy meadow plants and succulents), mostly on public buildings, with the plan to add an extra 10,000 square metres (12,000 square yards) in 2018. In 2016 the environment minister Rafael Pacchiano announced they would be planting an extra 18 million trees in Mexico City in what he called a 'historic reforestation' to reinforce the 'green belt in the megalopolis'.

While it makes intuitive sense that living near greenery means breathing better air, there is plenty of science to back it up too. A study by the University of Lancaster published in 2012 found that effective planting of vegetation in cities can reduce street-level concentrations of NO_2 by 40 per cent and PM10 by 60 per cent. Trees, as we have seen, are big emitters of VOCs, but their uptake of 0_3 and NO_2 and ability to filter our PM pollution, more than make up for it. The Lancaster researchers concluded that the 'judicious use of vegetation can create an efficient urban pollutant filter, yielding rapid and sustained improvements in street-level air quality in dense urban areas.' PM is removed by plants through a process known as 'dry deposition', whereby particles get stuck to the

---

[*] Compared to $3m^2$ in Tokyo or $6.4m^2$ in Istanbul. The WHO actually recommends a minimum of $9m^2$ per inhabitant, while the lucky citizens of Vienna, Austria, luxuriate in their $120m^2$ each.

leaf wax and are later washed away by rain. Another University of Lancaster study in 2009 found that a single tree could lower PM10 concentrations around it by 15 per cent. The gases, meanwhile – predominantly CO_2, but also NO_2, O_3 and sulphur dioxide – are sucked into the leaf stomata, the tiny openings or pores that plants use to breathe.

Simply planting more trees are easy starting points for cities wanting to clear their air. But some are already thinking on a far more ambitious scale. Stefano Boeri is a Milan-based architect internationally known for his 'vertical forest' designs. It's an idea that has long been talked of in sustainability circles, but few have actually gone ahead and done it. Stefano has. The Bosco Verticale (Vertical Forest) is a pair of 27-floor residential towers in the Porta Nuova district of Milan. Officially opened in October 2014, the exteriors are planted with over 700 trees, 5,000 shrubs and 15,000 perennials and climbing plants; the buildings take up just 1,500m² of ground space, yet provide the neighbourhood with the equivalent of 20,000m² of forest and undergrowth. This is not just a few extra pots on balconies – the planting and irrigation have been integrated into the infrastructure and design of the building. Trees and shrubs – the largest single one when planted was 9 metres (30ft) high, weighing 820kg (1,800lb), including soil – are irrigated with groundwater pulled up by a solar-powered pump system. In the hot Italian summer, heat inside the building is reduced by up to 30 degrees centigrade purely due to the vegetation, eliminating the need for air conditioning, while many species of birds have nested among the tree branches.

'The vertical forest concept was an idea that came to me in 2006,' Stefano tells me, still overflowing with enthusiasm for the project. 'I was pitching a project in Dubai, with more than 200 apartments in a building, and it was all covered in glass. I remember at that moment saying "this is crazy – in a desert!" I probably came up then with this idea of trees or forests and in that very moment … I suggested totally covering this project with trees or biological architecture.' The client turned it down, but his idea only grew. Stefano's mother, the designer Cini Boeri, was best known for designing a small house in the forest in the north of Milan – 'so I was always destined to work with trees in some way!' he laughs. When work finally began on the Milan towers, 'We studied many things that an architect never really gets to do, like how a tree can live at more than 100 metres, in terms of humidity, in terms of sunlight exposure, wind exposure, which kind of soil, how we can fix the roots to the base,[*] how to design the structure to cope with the weight of the soil, how we can propose the best maintenance system for the future of the building ... so we work a lot on that ... I came back to the client, Mr Heinz, who said OK, let's start.' The architectural design, says Stefano, started with the trees and plants: 'We designed the building according to the proportion and dimension of the trees ... We had a nursery first, of maybe 1,000 trees. We had to teach the roots how to grow into the spaces.' The growing process began in 2010, and 'we started to move the trees one by one from the end of 2012. The trees have been there now five years ... I consider it an open experiment. And what we are now doing all over the world, it's always referring to what we have done in Milan.' I ask what they have discovered five years on, which plants didn't do so well? 'All our plants are still there,' he says, 'which is amazing, we thought that maybe 10 per cent of the plants would die or suffer. But we had 21,000 plants when we started ... I think we had trouble with eight plants. That's it ... CO_2 is an amazing fertiliser! It is true. We are seeing this in China. When you plant trees in such harmful air pollution conditions, you see trees growing so well, so fast. It was something we were not expecting, but is also something you should consider as an opportunity, no?'

---

[*] All the trees have elastic bands that connect the root ball to a steel mesh embedded in the soil; all the medium and large trees also have a safety cable to prevent the tree from falling in case the trunk breaks; while the largest trees in locations most exposed to wind have a safety steel cage that restrains the root ball and prevents it from overturning in major windstorms.

His firm, Stefano Boeri Architetti, has since built vertical forests in Toronto, New York, Nanjing, Utrecht and Mexico City. But his grandest project to date is the Liuzhou Forest City, in southern China. Commissioned by the Liuzhou Municipality Urban Planning Bureau, it will be an entire forest city of 30,000 people, 40,000 trees and almost 1 million plants. With every wall and roof planted, it hopes to absorb almost 10,000 tonnes of CO_2 and 57 tonnes of pollutants including NO_2 and PM per year and produce approximately 900 tonnes of oxygen. 'In China we are looking at a totally different selection of trees to Italy,' says Stefano. 'In every place the architecture is a consequence of the selection of the trees ... First, we have to increase the numbers of parks and gardens, we have to work on reforestation, to surround cities with forests.'

The greening of buildings and architecture is a significant part of the answer to urban air pollution. 'We know that 75 per cent of CO_2 present in the atmosphere is pollution caused by cities,' says Stefano. 'And we know that 30 per cent of CO_2 is absorbed by forests all over the world. So, the idea to move forest inside the cities, in a way is to fight the enemy within ... it is not expensive ... I think it is a very, very *easy* way to fight air pollution.' Given that a one-hectare forest typically has around 350 trees, Liuzhou Forest City will be equivalent to a 114-acre forest, despite taking up far less ground space than that. No one yet knows, because it isn't yet built, but there is a very real possibility that – if combined with electric and active transport, and renewable energy – this could be the first carbon-positive city, that gives back more to the environment than it takes.

Plants and trees are an easy retro-fit option for older cities too. In France, a new law since March 2015 has required all new buildings in the country's commercial zones – shops, offices and restaurants – to have either solar panels or green roofs. This will provide habitats for birds, absorb PM and NO_2, retain rainwater, and provide insulation both in winter and summer. In Toronto, a similar law has required all commercial and large residential buildings built since 2009 to have at least 20 per cent green-roof coverage. In Zurich and Copenhagen all new flat roofs, both private and public, must be green roofs. And since 2001 in Tokyo, all new buildings larger than about 3,300m² (11,000ft²) are required to have at least 20 per cent usable green-roof space.

Singapore, a city-state with one of the highest population densities in the world – around 5.6 million people crammed into 680 square kilometres (260 square miles) – is trying to cut its annual PM2.5 by 12µg/m³ by 2020. Part of its plan to do so, despite obvious space limitations, is to provide 0.8 hectares of green space per 1,000 people by 2030. According to the Singapore Center for Liveable Cities, the green cover in Singapore was around 36 per cent in the 1980s; by 2016 it was up to 47 per cent, despite the population on the island having more than doubled in the meantime. All new structures in the city must include green roofs or green walls. A few hundred kilometres of cycling and walking trails now weave through the island, connecting a network of green spaces and waterways together, with the hope of spawning a cycling culture in the once car-dominated city. The ultimate goal is 640km (400 miles) of walking and cycling paths. But it's the 'supertrees' at the 250-acre Gardens by the Bay, looking like a utopian other-world from *Star Trek*, that really catch the eye – and the imagination. The artificial tree structures ranging from 24 to 48 metres (80 to 160 feet) collect enough solar energy to light up at night, their 'trunks' providing vertical gardens, with more than 150,000 plants weaving in and out of the wire branch-like frames.

There is something personally empowering about the greening of cities, too. If a single tree can lower PM10 by 15 per cent, then planting trees in front of your house would reduce the PM and NO_2 levels in your garden and in your home. If your local children's school or nursery is near a busy road – which is highly likely, given that most were originally built for easy access to cars and buses – then planting green walls with ivy or tall, dense evergreen shrubs can be instantly beneficial for the health of the kids who play in the playground. Their lungs, stunted if exposed to high levels of traffic-derived NO_2, PM and nanoparticles, will be protected by this additional green barrier between them and the pollution source. Recent research carried out by King's College London monitored NO_2 concentrations either side of an ivy screen installed at Bowes Primary School in north London, close to the busy North Circular road. A 12-metre (40-foot) screen of ivy was installed at the primary school, with air quality measurements taken during the months before the green wall went up and the months immediately afterwards. The ivy wall reduced exposure to NO_2 by nearly a quarter (22 per cent).

To highlight an air pollution problem, however, whether it's by a school or on your street, you first need to measure it. And most streets and schools aren't lucky enough to have a municipal air quality station on the doorstep. Kamila Knapp, co-organiser of Krakow-based Smogathon, tells me in 2017 that in Poland, 'there are only 126 of the huge official measuring stations, in the whole of the country – that's one for every 125 square kilometres [50 square miles].'* Smogathon runs a competition each year to offer seed-funding to smog-busting start-up ideas, and helped previous winners Airly.eu set up cheap sensors – about 150 sensors in Krakow alone – to give residents readings that are more relevant to their street and their commute. 'Now we see exactly where the air is bad and what needs to be done', adds Kamila's co-organiser, Maciej. 'If you go to Airly.eu … you can see exactly where the air is bad, and that's pretty cool, because you can fight it. They [the public authorities] were denying it for a while, because the problem is they did not officially measure it … the accuracy of Airly is over 95 per cent and they cost 200 quid – and the station costs 200,000 quid … So, the idea is that if you have a dense network, even if you have some anomalies and some sensors that don't exactly work, you can still average to a close accuracy … No one looks at the official station readings in Krakow now, they use Airly!'

---

* Which is better coverage than in the UK, by the way.

And, of course, there's consumer monitors like my Egg. In Beijing, I take the subway to Beixinqiao on the pink line, to treat my Egg to a homecoming. I'm on my way to Kaiterra, the makers of the Laser Egg 2. When I get off in Beixinqiao, I realise this is quite a different area to the Central Business District. Suddenly buildings are only one or two storeys high, the telephone cables and power lines hang low and loose, and narrow streets look more like back alleys. I take a street that is barely a few shoulder-widths apart. Red lanterns hang invitingly from restaurants. The Kaiterra HQ is almost indistinguishable from the other single-storey buildings that line this street, each one painted battleship grey, with curved roof tiles populated by moss and fallen leaves. A pagoda-like red arch leads to a small central courtyard. I only know I'm in the right place when I see that the yard is full of Kaiterra-branded boxes and products ready to package and ship. The windows around the courtyard are lit up in the late afternoon gloom by 20 or so people working at their laptops. A door opens and Liam Bates, the athletic, young CEO, dressed in T-shirt and jeans, walks out with a long, loping gait and greets me. In the little meeting room where we chat and drink green tea, bookshelves loom with titles by Elon Musk, Michael Bloomberg, and entrepreneurial self-help books with titles like *Getting Things Done* and *Radical Candour.*

Liam started out as an entrepreneur when he was just 12 years old. While his friends were earning money babysitting, he was designing websites – 'when you're 12 you can't get any real work, but over the internet no one knows.' He had to use his parents' credit card, but rather than taking money off it he added money ('they were very supportive'). He first came to China aged just 16 and founded a website helping foreigners come to study martial arts; he employed five people when still in his teens. Stints at university and TV presenting followed, until the Airpocalypse hit. 'I was completely oblivious to the problem of air pollution until the Airpocalypse – the sky went almost black during the day,' he tells me. 'My fiancée had a hard time breathing. When she was young she had asthma, but we thought it had disappeared long ago … She was like, we need to buy an air purifier. Then, I had this crazy idea – wouldn't it be fun to try and make one?' His first product, an air purifier called the OxyBox, came out in 2014. 'But then we realised that if you don't measure your air, purifying it properly is virtually impossible, and thus monitors are incredibly important … Before the Laser Egg, there weren't really any other low-cost air quality monitors available.'

I take my now bashed and bruised Egg out of my bag and reintroduce it to its maker. 'Ha, that's funny!' he laughs – he's previously only ever shipped them out, making this one the first to return home. I ask him to talk me through how it works, starting with the soft hum as it sucks air through the front? 'It is actually taking it from the back. It is pulled down a tube on the inside and a PCB [printed circuit] electrical board which has the convenient advantage of warming the air up slightly, which decreases the humidity level, which makes the readings more accurate … Particles cross the laser beam – hence the name "laser" egg – that causes the light to diffract, and underneath there is a sensor which picks up on the intensity of the light. So technically this device could and should be called a particle counter. What it is doing is counting every single particle which crosses that laser beam. And as it crosses it, it sizes it. Particles of different sizes cause the light to diffract differently … and it comes up with this number [PM2.5μg/m^3].' Laser Eggs have also been designed to communicate with each other, constantly receiving new calibrations according to their location. In major cities such as Beijing and Delhi, Kaiterra have set up fixed outdoor Eggs for this purpose. Hundreds of them. 'We are about to ship a couple of hundred more monitors to install across Delhi,' he tells me. 'So we will probably have ten times more data on Delhi than the Indian government. Which is ridiculous!' He types at an implausibly fast speed on his laptop and a graph emerges which resembles a foreground of softly undulating light-blue hills set against huge purple mountain peaks. 'So, this is insane,' he says, genuinely taken aback by what he is looking at. 'The light blue is Beijing's air quality, and the purple is Delhi, for the last two months [November and

December 2017]. There were maybe two days when Beijing had worse air than Delhi … Actually no, that's just *two hours*.' It shows Delhi's November peak, when the official PM2.5 sensors couldn't go past their three-figure settings of 999μg/m^3 – here we can see that a Laser Egg located beside the US Embassy in Delhi actually reached 1,486μg/m^3.

Where consumer particle counters such as the Egg fall short, however, is their inability to register nanoparticles. It can only register particles down to 300 nanometres, or PM0.3. After that, the particles are too small for the laser to pick up. I get the sense that Liam's faith in PM2.5 creates a blind spot for ultrafines. He tells me that he believes the pollution in Beijing 'is industrial – steel production is a huge part of it – and power generation, which is coming from coal,' meaning that, 'banning cars in Beijing will have negligible effects.' To back up the point, he tells me, 'If you take the Laser Egg on the street in Beijing and stand in the centre of the third ring road, with cars everywhere, there is no change in the PM2.5 readings, because they are so far beneath the background level of PM2.5. Even if you go to a car's tailpipe, the air coming out is actually no worse than the air in the city! In fact I have actually seen the reverse happen, where you put it at a car's tailpipe and the numbers go down, because they have a good filtration system.' Transboundary industrial pollution is the cause of most of Beijing's PM2.5 woes if your cut-off point is PM0.3 or 300nm. But given that studies of ultrafine particle emissions from road vehicles have found that 90–99 per cent of PM2.5 (by number, not by mass) are smaller than 300nm, that's a pretty big blind spot. And given that the Edinburgh gold study found that you have to go right down to 30nm before you find the deadliest particles that can enter the bloodstream, those are the ones we should care about the most.

What the Egg does very well is to give an indication of personal exposure in someone's home, street or commute, far better than a single fixed monitor ever could – even the precise nanoparticle readings of government monitoring stations are only helpful if you are standing right next to one. But while the Egg is cute and portable, it couldn't claim to be wearable technology, like a FitBit or Apple Watch. In Paris, I meet a low-cost sensor entrepreneur who is trying to produce the world's first small, wearable device that can clip onto a belt or bag and measure real-time concentrations of NO_2, VOC, PM2.5 and PM10. In the offices of his company, Plume, Romain Lacombe talks at a breathless, caffeine-fuelled pace. The pre-orders for 'Flow' – his stylish wearable air pollution monitor the size of a cigarette lighter, in charcoal grey, set into a mock-leather holster – have just opened, and orders are coming in fast from across the globe. Everything about the Plume offices screams 'modern tech start-up'. Casual clothing, coffee machines, technicians working shoulder-to-shoulder behind unnecessarily large monitors. There is even the obligatory table tennis room.* For the past three years, Plume's main business has been in air pollution forecasts – specifically the 'Plume Air Report' app, which offers a two-day pollution forecast, including minute-by-minute levels. It started in a handful of major world cities, simply using publicly available monitoring data; then 60 cities, then 430. In late 2017 it began to use satellite data and atmospheric modelling to fill in the gaps, and now even covers small towns like my own in Oxfordshire. 'The Air Report app was a way for us to start building a user base and explore how we could show that people could actually change their habits,' he explains. His plans for Flow, meanwhile, are more ambitious: 'Even with very precise on-site monitoring, personal exposure actually varies quite a bit based on where people live, the type of activities they do, how much you ventilate or not. So, if you go running in an area with high pollution and you are hyperventilating, of course you are going to breathe more pollution than someone else … The reason why environmental agencies invest so much in having fixed monitoring

* Note for all business historians of the future – all 'tech start-ups' in the early 2000s had to have a 'fussball table'; in the 2010s, it was replaced by table tennis.

infrastructure is in order to determine when action should be taken. But what's not measured is what people actually breathe in their day-to-day lives, when they go to work taking the subway or bus or driving or biking behind a truck, when they are at home … what we're trying to do is focus on that personal exposure.'

When I ask how Flow compares to official city fixed monitors in terms of accuracy, Tyler Knowlton, Plume's Canadian director of communications, steps in: 'these aren't designed to replace those fixed monitors – you kind of need both … These don't test up to EPA (US Environmental Protection Agency) fixed point standards, that's not the point. But it's like the canary in the coal mine. This can trigger action. If you can start to see that your exposure is too high, then you can investigate that further … I have a friend at the EPA who works on the massive, millions-of-dollars monitors – it's like comparing a home telescope to NASA. But I feel like the way we as a society are approaching this is totally busted – because those fixed points, even if you are lucky, there's only a dozen of them in your city. If you live in a small town in Michigan, there's zero, and even if there is one there's little capacity to collect and understand the data. Pollution doesn't just sit and hang around politely near the fixed-point station, and then replicate itself out uniformly across a city. It moves and changes and shifts, it is up and down and all over the place. You need both – the big things to take an average, but also a system that is always in flux, like the air you are trying to measure.'

Even the much-maligned (by me and Jim Mills, in Chapter 2) plastic diffusion tube, which is only able to give an average NO_2 level, is becoming popular among concerned citizens. The environmental charity Friends of the Earth began a campaign in 2016 for the public to buy a 'Clean Air Kit' from its website for as little as £10 ($12), which included a diffusion tube, instructions on how to install it, and a Jiffy bag to send it back for free analysis. To date thousands of Clean Air Kits have been sent out, with the results pinned to an online map of the UK. And despite being an appallingly blunt tool to measure NO_2

with, in terms of evidential quality to take to a local authority, it is actually far better than the live reading offered by the likes of Flow and the Laser Egg. Jim Mills tells me, 'if you went to your local authority now and said … I've used a Chinese particle monitor and got this reading, I'm sure they would be very kind and they'd listen. But then probably after the phone call then they'd say "there's another guy buying a cheap device and telling us that our monitors are all wrong".' But surely a diffusion tube is far less accurate, I ask? 'Yes. But if you buy a ten-quid diffusion tube and stick it up for a month and you send it off to the lab and it comes back over the statutory limit then you have every reason to insist that they do something about it, because that information, even though it's just as uncertain [as an electronic device], is credible because we know what the uncertainty is and there's evidence over decades.'

I decided to put some of the theories in this chapter to the test. My daughter's nursery stands next to a roundabout on busy road. During rush hour, four queues of traffic trudge slowly forward, with the nursery building at the epicentre. Immediately behind some flimsy wooden gates is my daughter's playground. I contacted the nursery to ask if they'd let me put up some diffusion tubes to test for exposure levels, and potentially erect a green wall to replace the flimsy fence between the playground and the road. I wasn't sure how open they would be about discussing this issue – most schools in the UK are government-run, but most pre-school nurseries for working parents are for-profit, private enterprises. Negative associations such as high air pollution would be bad for business. My request would cost time and money, both of which are in short supply within the childcare sector. Fortunately, the director of the nursery was keen to meet me.

I arrive for the meeting armed with reports on air pollution, council air quality readings, and brochures on the benefits of green walls. The director, Tom, is an English eccentric, sporting an out-of-season Father Christmas look with a bushy white beard and a woolly green waistcoat. He quickly says yes to my idea of putting up some diffusion tubes and says he will immediately look into building a green wall. I order a Clean

Air Kit from Friends of the Earth and put two diffusion tubes up in the roadside playground for two weeks. The results come back a couple of months later showing the playground had an average NO_2 reading of $20.4\mu g/m^3$, while one I placed on the building furthest away from the road recorded just $1.3\mu g/m^3$. Despite being well below the EU average limit of $40\mu g/m^3$ (the point at which, Jim Mills suggests, I could have demanded some action from the council), the roadside reading is still a concern – if the average is that high, including at night when there's no traffic, then peak rush-hour readings must be much higher. I suggest going ahead with a green wall, and, from the research literature, recommend a thick tall barrier of evergreens such as conifer. The pollution literally sticks to the leaves, making deciduous trees useful for only half the year, I tell Tom. 'Yes, yes, good – simple, that's how I like it. Whatever's best for the children and for your daughter,' he says. I leave wondering why I worried. A school or nursery should always have the children's interests at heart.

Weeks later, the planters arrive, along with 11 young conifer trees, already 2m (6ft) tall. Their immediate impact will be limited, but within a year or so they will grow taller and denser, and my daughter's playground will have lower concentrations of pollutants than it had before, making it safer for her and her friends to play outside. I feel happier, relieved, maybe even a little proud. But I still find myself glaring at the daily queues of cars pumping out their fumes every morning when I take her in, and every afternoon when I pick her up. Without them, there would be no NO_2 problem to solve.

CHAPTER TEN

What Price Fresh Air?

I'd been looking forward to my interview with Dr Devra Davis, a Donora survivor and one of America's leading epidemiologists. But, I must admit, I hadn't thought to research her husband. On the phone she mentions that they have travelled together to Michigan to see their grandson perform a Brahms piano concerto, and I can hear the clattering of bowls and cutlery as someone prepares breakfast in the background. This call has been in the diary for weeks, but I have a feeling that – given the purposeful clamour of cutlery – whoever the 'someone' is, isn't very pleased about it.

By the middle of the interview with Devra, whose 2002 book *When Smoke Ran Like Water* did much to highlight the issue of air pollution, we have discussed the epidemiological and toxicological evidence. Then we get on to the lack of urgency among policy-makers. Why is so little being done about an issue that destroys – and ends – so many lives? 'That is one of the reasons why I continue to do this work at my advanced age,' she sighs. 'The whole approach to environmental policy rests upon a very fundamental requirement, which is: show me the bodies. What the economists say is, until you have dead bodies, we don't have proof.' What happened next in our conversation unfolded like a stage play. So, picture the scene: a married couple, an epidemiologist and an economist, are on vacation, trying to prepare some breakfast, when a journalist calls …

> *Devra* [testily, on the phone]: My husband is in the room,
> I should say.
> *Man* [bangs the cereal down, pointedly, on the table]:
> I would like to make a statement.

Devra: OK yes, please do! My husband is an economist, and by the way one of the chief economists of the EPA. We've had long-standing discussions on exactly this issue for 40-something years. So, this is Richard Morgenstern, former senior economist and acting deputy administrator of the EPA. [She hands him the phone.]

Richard: Devra points to situations which have turned out historically to support her point of view, no question about it. However, not all problems which are identified in initial studies turn out to be as severe as, say, [Tetraethyl] lead or something like that. So, what you have is a situation where we are all dealing with unknowns, society is dealing with unknowns … we operate in a world of uncertainty. Now, there are cases where people believed that certain chemicals were really very harmful, and it turns out that they are not as harmful to humans as they were in the animals that were tested – the one I'm thinking of was a component of gasoline that was tested in a certain type of animal, and it turned out that the carcinogenic effects found in the first animal were not supported in other animals. But the larger point is, you have to take into account the magnitude of the change you are asking for. It seems simple: you do a study, you find a problem, therefore you should ban it or try to reduce the exposure. But sometimes it's not that simple. And sometimes communities actually want the jobs and they understand the risks.

Devra [shouts in the background]: Tell him about those coke oven workers in China!

Richard: I will! Many years ago I was in China – I have been there many times, but this was one of the earliest times I was there – there were people on top of a coke oven battery sweeping away the materials. Now we [the US] banned that a long time ago, we don't allow people to do that, and we introduced machines to replace that. The Chinese were still using people when we were there. So, I said to the person who was guiding me through the steel plant, 'Gee, do people know that that is extremely dangerous – it has been documented that benzene emissions from coke ovens are real killers.' And he said 'Oh yes, we know that.' So I said, 'Do the people up there working on the coke ovens know that?' And he said 'Oh yeah, they know that.' I said, 'So why are they doing it?' And he said, 'Well, they are getting a substantial increase in pay compared to other workers in the plant, and they are voluntarily accepting these risks.' That brings me back to the economic and social question: yes, we have research that shows problems, but people react to them very differently – some people go for the jobs, they prefer the income, they prefer what they perceive to be the benefits of having these problems. In this case, there was no uncertainty because the scientific evidence was overwhelming.

Devra: Scientific evidence meaning we have plenty of proof of sick and dead bodies ...

Richard: ... Yes that's right, associated with benzene from coke ovens, that was established. But in some of the other examples out there, there is uncertainty and it doesn't always work out to be such a problem. But people say 'yeah, I want the economic gain'. So, the dilemma for people in policy positions is to ferret out the situations where the economic gain is either very small, or the health damages are very large – those are the things that we regulate. But it's a tough world out there.

Tim [by this point feeling less like a journalist and more like an interloper in a marital dispute]: But isn't that also a common, consistent delaying tactic from industry and politicians, to say that 'unless 100 per cent causation is proven, then more research is needed before we do anything'?*

Richard: Yes, it is a delaying tactic, I'm not disputing that. I'm just saying, if you look at it narrowly from the health perspective and the adverse effect on people, then it is just delay. If you look at it from the social perspective of 'is there value to society of having people do this type of work, and what are our other options?' Think about World War Two – all the people in Britain and the US worked in very hazardous jobs producing munitions and armaments, during which many of them died – forget the warfare itself, they died in the production process. Was that worth it? Well, as a society, most people would argue that the alternative was worse. There are a lot of issues here, and I think it is a little simplistic to focus narrowly on some particular losers in the system, because there are some complicated ...

Devra [exasperated]: Oh God!

Richard: Well anyway, that's my story.

Devra [to Tim]: So you can see why we keep having what we call 'intense dialogues'?

Tim: So, what's your comeback to what he just said, Dr Davis?

Richard: She doesn't have one! OK, let's hear it, let's hear it!

Devra: It's a question of the downside risk.

---

* The erstwhile head of the EPA in 2017, Scott Pruitt, was a big fan of this tactic. In mid-2017 he made a statement saying 'what the American people deserve ... is a true, legitimate, peer-reviewed, objective, transparent discussion about CO_2' – the inference being that such an exercise hasn't happened, despite in reality having been ongoing in thousands of peer-reviewed journals for decades, including five reports from the Intergovernmental Panel on Climate Change (IPCC). In late 2017, Pruitt again told CNBC, the American news channel, that there is 'tremendous disagreement' about whether man-made emissions are causing climate change, and that 'we need to continue to debate, continue the review and analyse'. There is very little disagreement among the scientific community on this issue. However, by repeatedly saying there is, the status quo can be maintained indefinitely.

Richard: Right!

Devra: Do you value more the value of human life and the quality of life, or do you value more the immediate economic progress? And as John Maynard Keynes once famously quipped, 'In the long run, we're all dead.' Of course, economists don't tend to think about the long run, they think about the short run. Now I don't want to put anybody out of work, but I think that we tend to dismiss health issues, especially when there is an immediate and obvious short-term gain … As an example, the Swedes – who obviously have a much more homogeneous society than we do – they managed back in the 1970s and 1980s to cancel about 80 per cent of pesticides because the risks – many of which were not proven but just suspected based on experimental evidence – were such that they did not want to expose them on their citizens. So they just cancelled them. We in the US continue to use some of those same pesticides today, because the industry here, especially now, is so much more powerful than those of us concerned about health issues.

Richard: So are we less healthy than the Swedes, as a result?

Devra: On some of these issues yes, as a result. In terms of Alzheimer's and dementia, and …

Richard: … But you can't demonstrate it. You can't demonstrate it's not due to other differences between the US and Sweden …

Devra: Excuse me. I stipulated that in my prior response. They are a more homogenous society than we.

Richard: My final comment is, that it's one thing for a – and I won't say that Devra is any more upper middle class than I am – but for either of us who are relatively comfortable in society to offer such advice to people who are struggling to make a living. And that's the sort of trade-off, that's the issue – a lot of the people who are placing health protection above economic cost are themselves quite protected.

Tim: But it doesn't have to be an either/or though surely – isn't it about protecting the most vulnerable?

Devra:Yes, it doesn't have to be an either/or …

Richard: It is true that there are some free lunches out there,
   but it turns out not that many. Most of these things
   come with a cost, whether obvious or not so obvious.
   Look, I am in favour, I am not anti-environment at
   all, I'm just pointing out that it's a little simplistic to
   ignore the economic impact.

I loved every second of that conversation. Because – and I
couldn't help but think this while they were talking – it
highlights a crucial 'final chapter' point. Which is, despite
everything we've learned about air pollution, can we afford
to change, or should we just learn to accept it? Is the economic
impact of clearing the air a net loss, or a net gain?

Given that air pollution is first and foremost a public health
issue, let's answer that question by looking at the public health
costs and benefits first. According to the *Lancet* Commission,
treating air pollution-related diseases is responsible for 1.7
per cent of annual health spending in high-income countries
and up to 7 per cent of health spending in middle-income
countries. 'The costs attributed to pollution-related disease
will probably increase as additional associations between
pollution and disease are identified,' the Commission found,
referring to the new disease links to air pollution we're
discovering seemingly every day. The OECD believes that
global air pollution-related healthcare costs for its member
countries will increase from \$21 billion (£16 billion) in 2015
to \$176 billion (£134 billion) in 2060, while welfare costs
would be well into the trillions. United Nations figures in
2016 show that air pollution across Europe is already costing
\$1.6 trillion (£1.2 trillion) a year in deaths and diseases,
nearly one-tenth of Europe's gross domestic product (GDP);
in 10 European countries, the cost was above 20 per cent of
GDP. A South African study in 2012 found that 7.4 per cent
of *all* deaths that year were due to chronic exposure to fine
PM, costing the country up to 6 per cent of its GDP, or \$20
billion (£15 billion). For Africa as a whole, the estimated
economic cost of premature air pollution deaths in 2013 was

roughly $215 billion (£175 billion) a year for outdoor air pollution, and $232 billion (£177 billion) for indoor air pollution.

I could go on with stats like that for the rest of the chapter (though for the sake of mercy and readability, I won't). The total economic cost of air pollution in London was £3.7 billion in 2010 (sorry, last one). But where exactly do such costs come from? To give an example, in the journal *Environmental Health Perspectives*, a 2015 paper led by the New York University School of Medicine calculated the economic costs of pre-term births caused by PM exposure – just one health effect, from just one category of air pollutant. The costs included treatment for medical conditions caused by pre-term birth in the first five years of life, and costs accrued in the years afterwards due to associated developmental disability and lost economic productivity due to reduced cognitive ability. In total, annual pre-term birth-related costs attributed to PM2.5 were estimated at $5.09 billion (£3.88 billion) in the US. The findings, say the authors, 'provide a sense of the potential economic benefits that could be achieved by regulatory interventions aimed at reducing air pollution exposure during pregnancy'.[1]

A cost-benefit analysis by the EPA in 2015 explained that:

Avoiding incidences of premature mortality, especially those associated with exposure to fine particles, contributes the vast majority of the direct benefits of 1990 Clean Air Act programs measured in dollar value terms … First, the differences in air quality, human exposure, and resulting risk of premature mortality … are substantial. Second, these changes in risk of premature mortality are estimated to have significant economic value.

Welfare outcomes and worker productivity have a big impact on economies beyond the direct healthcare costs, too. The *Lancet* Commission found that productivity losses caused by pollution-related diseases reduce GDP in low-income to middle-income countries by up to 2 per cent per year.

Reduced workplace fitness and productivity globally due to pollution are estimated to amount to $4.6 trillion per year: 6.2 per cent of global economic output. To understand how this is possible, let's again zoom in to the micro scale. A 2016 study conducted by the Marshall School of Business of two call centres in China found that increases in the pollution levels found in the Air Quality Index (AQI) correlated with a decrease in the number of daily calls handled by the call centre operatives: workers were 5 to 6 per cent more productive when air pollution levels were rated as good (AQI of 0–50) versus when they were rated as unhealthy (AQI of 150–200).[2] A separate study of agricultural fruit pickers in California found that a 10ppb change in ozone resulted in a 5.5 per cent drop in worker productivity, measured in the amount of fruit they picked. The authors calculated that a 10ppb reduction in ozone across the US would therefore translate into an annual cost saving of approximately $700 million (£530 million) in agricultural labour.[3] All of which starts to turn the question 'can we afford to clear the air?' on its head – can we afford not to?

The UN estimates that the global benefits of phasing out leaded petrol alone amounted to $2.45 trillion (£1.87 trillion) a year in improved health, fewer early deaths, increased IQ (and therefore increased lifetime earnings) and decreased violent crime. Frank Kelly tells me that, in the European context, wherever measures have been introduced 'to decrease emissions from urban areas and remove coal power stations from urban areas, [governments have] seen an improvement in air quality, they have seen an improvement in life expectancy and decrease in certain diseases, which they can't explain by other causes … that if you do spend this money here, you are going to get a big benefit, because the health costs are always more.' As for the chosen means of reduction, such as replacing short car journeys with active travel such as walking and cycling, the UK Chief Medical Officer's annual report makes the point that – aside from the consumer saving on petrol and car maintenance – 'getting one more child to walk or cycle to school could pay back as much as £768 or £539 respectively in health benefits, NHS savings, productivity gains and reductions in air pollution and congestion.'

Veerabhadran Ramanathan, a professor of Atmospheric and Climate Science at the University of California, San Diego, wrote in a WHO news bulletin in 2016 that 'the tragedy is that there are perfectly feasible solutions to the air pollution problem, but these are surrounded by myths'. The biggest myth, he believes, is that tackling air pollution is more expensive than the benefits: 'In California we found that if you clean up the air, each dollar invested in air pollution returned nearly $30 to California. There were huge health benefits along with a large increase in new jobs and thus in people's well-being.'

Another big myth is that we need to dig up and burn fossil fuels in order to grow economies. Since 2010, GDP in Germany has been on an upwards trajectory while gross power consumption, primary energy consumption and greenhouse gas emissions have all been on a downward trend. At the Paris climate talks in 2015, the California Senate leader similarly boasted, 'We have successfully decoupled carbon from GDP.' The amount of petrol pumped in California has declined every year since 2009 while the state's economy grew by 5 per cent over the same period. Fossil fuels don't, in fact, come cheap. According to one report, Europe provided at least €112 billion (£99 billion) in subsidies per year between 2014 and 2016 towards the production and consumption of fossil fuels. We continue to fund coal companies, which are no longer profitable without government support, because they have entrenched and powerful political lobbies: in Europe, at least €3.3 billion (£2.9 billion) per year is provided annually in financial support to private (not state-owned) coal-mining companies alone. And that's just the domestic markets. Countries spend a huge amount importing coal and oil; remember that a third of all shipping is simply transporting oil around the globe. According to the US Central Intelligence Agency, the US is the world's biggest importer of oil at 7,850,000 barrels of oil a day, exporting just 590,900 barrels a day. India is the world's third largest importer of oil. Whichever way you look at it, the idea of becoming self-sufficient with renewable energy makes complete economic sense.

For anyone doubting the impact of government policy on this issue, consider this quote from Melba Pria, the Mexican ambassador I met in Delhi: 'Mexico imports 1.5 million cars from India per year – seemingly every taxi or Uber is a Vento. The 1.5 million cars we import from India have a catalytic converter, because those are the standards for us to buy a car, from anywhere, and the catalytic converter has to have X, Y and Z standards. So that technology is built in India. But they don't use it in India, because they don't have the same regulatory standard. They are also exporting petrol gasoline that doesn't have sulphur, aluminium or lead, but they are not using it.'* Another economic own goal, noted by a report to the Indian Supreme Court in 2014, was that buses paid much more in road tax than cars, contributing to the boom of private taxi operators at the expense of bus companies.†

Clean air legislation has historically been shown to work with dramatic effect. The UK's Clean Air Act (1956) dealt with the causes of air pollution of the day, most notably sulphur dioxide and PM10 from coal smoke, with astonishing speed. The Act granted local authorities the power to designate smoke control areas where only authorised smokeless fuels could be used, with 40 per cent grants made available to householders to replace coal fires with gas or electric. It was a carrot-and-stick approach, recognising both the upheaval it would cause householders as well as the urgency required. By the 1970s, the air in the UK's major cities had changed beyond what anyone had thought possible. We need to do the same again. The success story of diesel cars – rising from 22 per cent of new car sales in western Europe to 51 per cent in just nine years (from 1997 to 2006) – also shows just how quickly

---

* This kind of double standard is not unique to India. H. F. Wallis, writing in 1972, asked why the UK car manufacturers 'neglect to build safety factors into cars for the home market while compelled to do so for those meant for export'.

† Electric rickshaws were also briefly popular in Delhi before being banned in 2012 for supposed 'safety reasons'. After a few years in the courts, an 'E-Rickshaw Bill' was finally passed in March 2015, allowing electric rickshaws, albeit with restrictions on load and speed. Potential e-rickshaw drivers, however, were now understandably reluctant to make the switch.

change can happen with a shift in government policy. If the same was done to promote electric or hydrogen fuel-cell cars over diesel, then levels of NOx and nanoparticles – the worst pollutants to human health – would fall off a cliff.

We are starting to see the first signs of this happening. In October 2016, the Bundesrat, the German legislative body, passed a resolution to approve only emission-free cars for use on the roads by 2030. It also called for the European Commission to consider implementing the same across the entire European Union. Tokyo has effectively banned diesel engines since 2003. At the C40 Mayors Summit in Mexico City in December 2016, mayors of four major world cities – Paris, Madrid, Mexico City and Athens – committed to banning all diesel vehicles from their cities by 2025. They also committed to doing 'everything within their means' to incentivise the use of electric, hydrogen and hybrid vehicles, as well as improve the infrastructure for cycling and walking. The mayor of Paris, Anne Hidalgo, avoided any possible accusation of ambiguity by saying, bluntly, 'we want to ban diesel from our city'. This wasn't just empty rhetoric: her Paris Crit'Air scheme (see Chapter 7) has revealed a step-by-step means of doing so.

In agriculture, too – the number-one emitter of ammonia – emissions regulation actually gives back more to the farmers than it takes (and, quite obviously, for the society around them too). The Netherlands reduced ammonia emissions by 64 per cent between 1990 and 2016 through various regulations regarding manure treatment and an emissions certification for animal housing. The schemes cost the Netherlands an estimated €500 million (£440 million) a year, but resulted in annual societal benefits of €900–€3,700 million (£800–£3,280 million) including €150 million (£133 million) in fertiliser savings for farmers.

Even if you are convinced about the net economic returns from reducing air pollution, there is still the question: where does the money for such schemes and infrastructure projects come from in the first place? California's answer, described in Chapter 7, is to make the polluters pay for it. The proceeds of California's cap-and-trade system are distributed to the communities most affected by air pollution as grants or loans. Other cap-and-trade schemes around the world such as the EU's Emissions Trading System, and China's scheme, launched at the end of 2017, could do the same. Cap-and-trade was a central part of the US 1990 Clean Air Act amendment, brought in during Devra Davis's husband Richard Morgenstern's term at the EPA, to target acid rain and sulphur dioxide emissions. But even cities without anything as fancy as a cap-and-trade mechanism can – and undoubtedly already do – have some form of monetary fine for pollution offenders: this should be diverted towards clean transportation infrastructure and incentive payments for residents to switch to clean transport and heating options.

Such policies are never popular with industrial manu-facturers, of course. James Thornton, who founded the NRDC's Los Angeles office alongside Mary Nichols, argues, 'the car industry has always said "That's impossible, that's impossible, that's impossible" with every single regulation, and it turns out it's a lie every time … They have sometimes used lawsuits to slow down California, but … it's always very responsible regulation, it's always based on good science and that's what you need to do.' Since taking the fight to the UK, he has met similar opposition from the UK government in response to meeting EU NOx limits. The UK, largely forced by ClientEarth's legal action, finally announced some NOx and traffic reduction measures in 2017, including the ban on new diesel and petrol cars from 2040. But the government has been slower to move on other vital measures ClientEarth has long called for – including emissions charging zones and diesel scrappage schemes. The idea behind 'diesel scrappage' is to offer car drivers grants to trade in their old diesel cars and replace them with new electric ones. The UK government previously ran a vehicle scrappage scheme in 2009–10; in the midst of a recession, it offered the domestic car industry a much-needed lifeline to sell new models as well as helping to meet national emissions reduction targets. The grant of £1,000 ($1,400) towards the purchase of a new vehicle when a car or van over 10 years old was scrapped was matched by manufacturers, giving a total of £2,000 ($2,800). The same is needed now, for exactly the same reasons. Domestic electric car makers, such as Nissan, whose Leaf model is made in the north-east of England, need a new customer base, and cities have stricter NOx criteria to meet. Diesel cars need scrapping.*

'Then the question is, where will the government get the money to pay for that?' asks Thornton, helpfully. 'The Germans just got a contribution of a very large amount of money from the car industry, I think €250 million, towards such a scheme. What I suggested to Michael Gove [UK Environment Secretary] was that he bring a case against the car companies [that installed diesel defeat devices] and settle it for somewhere between £3 billion and £5 billion and that money then goes into a scrappage scheme.' What was his reaction? 'I suggested that we could do a legal memo showing him the laws he could use and he said "I'd be very happy to see that memo". So we'll see ... In America I think Volkswagen settled for $20 billion [£15 billion]. You don't need $20 billion to do a great scrappage scheme.'

The levers that governments can pull will always be greater than those available to business or civic action groups. When I visited the low NOx district heating boilers in Beijing, Manny told me, 'We met with a factory in Shandong, and they said "thank you but we don't need your technology right now because our NOx standard is 50ppm." They basically said "it's great you can get to 5ppm, but we don't need to" ... if the regulation says you have to get to 5ppm, then people will buy it. But don't believe that the guys out at the factory level are altruistic and inspired to get to a lower level ... the key to all this, what will drive decisions, is regulation.'

However, regulation is only as effective as its enforcement. When the Israeli Clean Air Act came into effect on January 2011, it required big industrial plants to obtain an emission permit from the Ministry of Environmental Protection, with

---

* By late 2017, Nissan gave up waiting for the government and began to offer its own scrappage scheme, offering UK customers up to £2,000 ($2,800) plus trade-in value on their old vehicles to switch to a Nissan Leaf. But it was only a one-month trial.

limit requirements set for each industrial plant. Ronit Piso, formerly the Director of the Coalition for Public Health and a leading health campaigner in her home city of Haifa, believes that it has not always been enforced: 'The Ministry of Environmental Protection have a permit to get into any factory in the area and do their own emissions test. The gatekeeper has to let them in. I was personally, with my own team, inside one of these visits … we went to an oil tank factory in Haifa Bay and the gatekeeper kept us for more than three hours at the gate, because the director refused to let us in. Completely against the law. The ministry did nothing about it. I said "hey, they deserve a fine, you should do a special hearing, you should prosecute the director" – nothing … We have the industry with lots of money, lots of networking with high-rank, high-profile people in government, they control the media, actually they control everything – and on the other side we have the citizens, who on a weekly basis put up photographs on their own websites of pipes that pollute the area, all kinds of gases that appear in the sky. They don't feel any confidence in the authorities any more. This is a very, very major issue.'

This was the major focus of Chai Jing's 2015 Chinese documentary *Under the Dome*, too. Enforcement of emissions legislation in China was shown to be surprisingly weak, with city mayors and business CEOs openly admitting on camera to flouting regulations. Following the outcry caused by the film, China swiftly stepped up its enforcement efforts. In 2016, China's Minister of Environmental Protection, Chen Jining, told state media that 1.77 million enterprises had subsequently been inspected, uncovering illegality at 191,000 of them, temporarily halting production at 34,000 and closing down 20,000 altogether.[*]

---

[*] China named Chen Jining as its new Environment Minister one day before the documentary was released. This adds weight to suggestions that *Under the Dome* was not a scandal that shocked the authorities, but was fully choreographed with their help. According to a report by the BBC, the documentary script and interview were sent to the National People's Congress for comments and feedback prior to its release.

In Delhi, according to the *Hindustan Times* in October 2017, despite the newly implemented Graded Response Action Plan (GRAP) to counter severe air pollution, which includes greater pollution control for vehicles, inspection of buses, trucks and interstate vehicles, stopping interstate diesel trucks, and if necessary an odd/even number-plate scheme – the enforcement wing of the transport department had only 250 employees for a city of some 10 million vehicles. Of those employees, fewer than 50 had the power to prosecute. Meanwhile the Delhi Pollution Control Committee, tasked with distributing green licences to commercial establishments, ensuring PM control at construction sites, stopping the use of thousands of illegal diesel generators, issuing public health alerts and monitoring the levels of pollution, reportedly had only 60 officials.[4] A new law which bans diesel cars over 10 years old and petrol cars over 15 years old is also hard to enforce, Dr Sharma at the Central Road Research Institute (CRRI), told me: 'Implementation now is the issue … and also corruption. Unlike in the UK and US where you can easily identify the vehicles which are 10 years old, in India where millions of vehicles are plying the Delhi roads, the monitors are not there. You are taking a risk driving a vehicle which is not permitted, but there is a very rare chance that you will be caught … it is a cumbersome process for the traffic policeman to get involved with these things, so what they will do instead of impounding they will simply *challah* – *challah* means monetary fine … Very good laws and orders have been passed, but enforcement is grossly lacking.' He does, however, have some personal sympathy for people who do flout the law. Dr Sharma himself owns a petrol car older than 15 years, 'in perfect working condition … it is meeting all the numbers, the only numbers it is not meeting is 15 years old. It is quite painful for me to take to a scrap dealer who is buying for 1 per cent.' Professor Khare at IIT repeats this sentiment too: 'this ban is not very acceptable amongst the public … when we buy a car, we keep them as a family member.' He also adds that stubble-burning in the regions surrounding Delhi is 'banned but not enforced'.

The contrast with Mexico City, and the radical steps taken there in its fight against air pollution, is stark. Ambassador Melba Pria tells me about going for a morning jog on a recent visit to Mexico City: 'I heard a tap tap of shoes running up behind me and suddenly a male voice saying, in English, "lady, lady – no run!" So I stopped. It was a policeman. And he said, "Are you American? Only Americans run when we are in [an air quality alert] contingency ... turn on your TV, madam," as if to say, don't be so stupid! I took out my mobile phone, and we were at [AQI]156. Contingency starts at 150 ... There is a public acceptance of what you can and cannot do.' A Phase I alert in Mexico City is triggered by an AQI of 150 (and/or an hourly ozone concentration of 155ppb or PM10 above 215µg/m^3 for 24 hours), meaning that schools and government institutions are expected to cease all outdoor activities, all public construction and repair work stops, and citizens – as Ambassador Pria discovered – are advised that exercising outdoors represents a health hazard. A vehicle sticker system similar to the Paris Crit'Air scheme also kicks in. All vehicles must have a holographic sticker on their windscreen, showing a rating of 00, 0, 1 or 2, with 00 being the lowest emissions (i.e. electric vehicles), and 2 being the most polluting. When a Phase I level alert is issued, vehicles with class 2 stickers are banned from driving, while class 1 stickers move to an alternate day 'odd–even' number plate system; during a Phase II (above AQI200), both 1s and 2s are banned. The zeros are exempt throughout. So, that's another blueprint point: no matter how good your clean air laws, you've got to back them up with rigid enforcement.

During the global clean-up, we've got to avoid unintended consequences this time. Diesel was promoted in the 2000s, possibly in good faith (though as we saw in Chapter 5, I have my doubts), to reduce CO_2 levels. In so doing, the resultant increase of NOx and PM – let's not sugar-coat this – killed tens of thousands of people. Similarly log burners were (and still are) incentivised for being a sustainable fuel source, irrespective of where the smoke ends up, leading to domestic stoves becoming a major pollution source in

Britain for the first time since the Second World War. Both those outcomes were foreseeable if policy-makers had spoken to a scientist or two. But there's an easy way of making it less likely to happen again: burning 'cleaner' fuels is not the answer. Let's not forget the meaning of 'renewable' in 'renewable energy'. The Earth's biological fuel supply is finite; its solar and wind energy is not (at least not in a timeframe that need concern human civilisation). Whether it's inside the engines that propel vehicles or for our heating and cooking, we need to stop burning fuels within densely populated areas. Some forms of renewable energy are benign in terms of air quality: solar, hydro, wind, fuel cells and ground- and air-sourced heat. Others, such as biomass, biodiesel and wood chips, cause PM and NOx emissions. So, it's pretty simple: in urban areas, choose from the former list, not the latter. In Brazil, a percentage of bio-ethanol content in blended petrol fuel has been mandatory for decades. There are many reasons why this is good, but air quality isn't one of them. A study found that ozone levels actually increase due to the added ethanol while PM2.5 levels are broadly the same as in conventional petrol. That is just one study, but it makes the point: burning fuel will always cause air pollution.

Electrification is the answer. When I visited Ally Lewis in York in summer 2017, he told me, 'the need to burn things for electricity has massively diminished now … I first taught a course on electricity generation 10 years ago when there was this tiny renewable fraction in the UK's energy mix and lots of nay-saying that it couldn't be integrated. Now 10 years later I can give lectures with the majority of the UK's electricity now coming from renewables. Wind is the largest resource today. There are days with zero coal. It's almost hard to imagine the rate of improvement in wind turbine gearboxes and [solar] photovoltaics … in the middle of the day five years ago gas-fired power picked up most of the slack, when demand increases, and you used to have this flat amount of nuclear, flat amount of coal … Now it's provided by solar. It's phenomenal.'

California's governor Jerry Brown told the One Planet Summit in December 2017 that California was on course to meet its 50 per cent renewable target a decade ahead of schedule. That, as Ally suggests, is thanks to the falling price and rising performance of the technology. According to Bloomberg New Energy Finance (BNEF), the price for solar PV between 2009 and 2018 fell by 77%, and for onshore wind by 38%. The cost of lithium-ion batteries for electric cars fell by 79% during the same period, a fact that BNEF's head of energy economics called 'chilling for the fossil fuel sector'. By 2018, more than 300,000 Americans were employed in the renewable energy sector, compared to just 52,000 working in the coal industry. This crosses party political lines. Traditionally Republican states such as Texas, Utah and North Carolina have each installed more than 1 gigawatt of solar (each able to power over 700,000 homes). California has since announced a renewable target of 100 per cent by 2045 and has called for the entire United States to go 100 per cent by 2050. If that sounds like wishful thinking, or just plain unaffordable, then consider the award-winning 2015 study led by Stanford professor Mark Jacobson: he looked at the feasibility of a 100 per cent wind, water and solar (WWS) power grid for the United States by 2050, and for a range of factors (such as baseload energy, how you store the energy and likely increases in extreme weather events) it found a future renewable grid to be 40 per cent cheaper than a conventional one.[5] It projected a renewable electric system cost of an average 10.6 cents per kWh, compared to a conventional grid of 27.6 cents per kWh.* The reason is simple, writes Jacobson: renewable energy 'requires zero fuel cost, whereas conventional fuel costs rise over time'.

That might be OK for the richest country in the world, but what about everyone else? In 2017, Jacobson *et al.* repeated

---

* Comparing the probability estimates – *i.e.* the range between best and worst case predictions - is useful here too, as even the worst-case scenario for renewables (14.1 cents per kWh) beats the best-case scenario for fossil fuels (17.2 cents per kWh).

the study for 139 countries and consistently found renewables to be more affordable than fossil fuel energy over time. They also argued that moving to 100 per cent renewables would create 24.3 million full-time jobs (net – accounting for jobs lost in the fossil fuel sector) as each country becomes responsible for producing and maintaining its own power supply, as opposed to outsourcing it to Saudi Arabia or Russia.[6] While certainly bullish, Jacobson and his numerous co-authors are hardly maverick outliers. A 2017 literature review of 24 papers modelling transitions to a 100 per cent renewable future found that, '100 per cent renewable systems have been shown in the literature to be not just feasible, but also cost-competitive with fossil-fuel based systems.'

Even Simon Bennett at the International Chamber of Shipping surprised me with a vision of what reduced shipping could mean for the world: 'Assuming the world does decarbonise, that's going to have an impact on the movement of fossil fuels – if nothing else, moving fossil fuels contributes to about 30 per cent of the demand for maritime transport … if the shore-based economy decarbonises, ships won't be required to move large quantities of crude oil around the world. So, the decarbonisation of the rest of the world is probably going to reduce shipping demand which will in turn reduce … emissions from shipping.'

You could argue, then, that we are already on the right path. That we can just sit back and wait for clean air to blow in with the coming winds of change. While there is an element of truth in that, there's a lot wrong with it too. For one, how confident are you that you'll be alive in 2050? How old will your children and grandchildren be? If we have decades of high NOx, PM and nanoparticles still to come, are you comfortable with the health consequences of that? Without public engagement and pressure, there's a wealth of vested interests happy to slow down the transition to clean energy indefinitely. And misguided politics can always screw things up. As just one small example, the UK energy department began a Capacity Market auction in 2014 for relatively small power providers to contribute electricity to the grid at times when the contribution of wind and solar is low. This caused a boom in the numbers of diesel generators. And diesel generators emit ... actually, we know this bit by now, don't we? The Capacity Market auction in December 2015 awarded contracts of over 1GW to small-scale diesel or gas generators. It raises the very real possibility that our future power supply will look much like the lone electric car-charging point I saw at the LCV show, hooked up to a diesel generator. The number of own goals inherent in a system like that are too many to contemplate.

Just as the world is waking up to the benefits of tackling air pollution, politics can just as easily derail it. The aftershocks of populism caused by the likes of Brexit and Trump threaten to turn the tide against environmental laws with the promise of returning to a rose-tinted industrial past. Although, as we have seen, the industrial past was anything but rosy. Devra Davies quotes a former zinc factory worker from Donora who remembered conditions so bad that, 'five guys had gone before me to shovel out the finished zinc. Each one of them keeled over, real sick ... I was the sixth one in. I couldn't take it either. I left. Spent a week in bed and never returned. Not many made it to the age of 30 as zinc workers.'

Does appalling pollution inevitably force the hand of politicians to clean up their cities? I asked the scientist and historian Peter Brimblecombe whether the UK's Clean Air Act of 1956 was an inevitable consequence of the Great Smog of London in 1952. 'No, I don't think it was inevitable,' he says. 'It was strongly resisted for a number of reasons. Harold Macmillan was very concerned, he was Minister of Housing during the smog ... his idea was that there were already, within the Public Health Act of 1936, plenty of smoke abatement clauses ... And many politicians worried enormously about how can we tell people what they do in their own homes? That this is an appalling affront to personal freedom.' It only won through due to political will and public support. 'For me the Clean Air Act was always important as a political statement about how you make environmental legislation,' says Peter. 'How you challenge personal freedom, how you provide new technology to do things, how you fund the generation of knowledge needed to improve the atmosphere.'

Then there are the car lobbies. I have some sympathy with those car companies who tried their best to meet increasingly stringent emissions standards set by the likes of CARB and the EU. But, as we have seen, many didn't try at all, and even when forced to, their models regularly failed when tested in real-world driving conditions. This still amounts to production lines full of models that need to be sold. In early 2018, the car industry body SMMT attempted a fight-back, with a marketing campaign promoting the environmental benefits of diesel cars (*déjà vu*, anyone?). Tweets and infographics released by SMMT included such golden nuggets as 'Latest Euro 6 vehicles are the cleanest in history' – which, in the context of the very dirty history of diesel, is like claiming a new half-pound cheeseburger is the 'healthiest ever' because it has a sugar-free bun. And that these cars 'feature technology that converts most of the NOx from the engine into harmless nitrogen and water before it reaches the exhaust' – my response to that being, according to the EQUA tests of Euro 6 models in real-world conditions (see Chapter 5), no they don't.

There are clean-ups on offer, and there are cover-ups. Historically, we are very delicately poised: we could get stuck in an indefinite cover-up phase. Cheap desk-top air filters for the home sell for as little as $20 (£15), but strain in vain against the volume of gases and particles we pump daily into the air. On my writing desk sits a voucher from a packet of washing detergent, inviting me to 'Keep calm and breathe deeply: Enter for a chance to win 1 of 8 air purifiers with this pack.' Breathable air is now a competition prize. In the 1980s sci-fi comedy *Spaceballs*, Mel Brooks's hapless President Skroob sucks down cans of 'Perrier Air' because the air on his planet has become so dirty. That sci-fi parody became modern-day reality in a worryingly short space of time when, in 2013, an enterprising Chinese businessman began selling cans of clean air on the streets. But that wasn't a one-off. In 2015, a Canadian start-up called Vitality Air shipped 4,000 eight-litre cans of 'Rocky Mountain air' to China for 100 yuan ($15, £11) a pop. In 2016, another Beijing entrepreneur started selling canned air – only this time it was of Beijing's polluted air, for tourists to take home as a souvenir. When I spoke to the shop owner in 2017, he told me he had sold thousands of them for around $4 (£3) a can, and 'I have a new pollution-related product hitting the shelves today: the Beijing Pollution Globe, which is like a snow globe with the iconic Beijing CCTV tower in and when you shake it it's covered in grey pollution floating around.'

There are grand plans to fight the symptom rather than the cause, too. When Dutch artist/inventor Daan Roosegaarde first visited Beijing in 2013, he came up with the idea of an outdoor air purifier – a giant vacuum cleaner that could suck PM out of the air. Three years and several prototypes later, Roosegaarde unveiled a 7m-high (23ft) 'Smog Free Tower', supported by the Chinese Ministry of Environmental Protection, in Beijing's 751 D-Park. It uses positive ionisation, a similar principle to static electricity, to attract airborne particles to it. 'We found a technology which is used indoors for hospitals and we scaled it up,' says Roosegaarde. 'We charge the small particles with positive ions, then a large negatively charged surface [inside the tower] attracts them … basically you suck the polluted particles in from the top, and you drag them down.' It captures and collects more than 75 per cent of PM in an area the size of a football field, running on just 1,400 watts, less than the average desktop indoor air filter. As for what happens with the collected PM waste once collected, he currently has a sideline selling the compacted black substance as jewellery. Prince Charles owns a set of 'smog-free' cufflinks. Roosegaarde and his team are now looking at how to scale up the Smog Free Tower: 'We're working now on the calculation, how many towers do we actually need to place in a city like Beijing to get a pollution reduction of 20 to 40 per cent? … It shouldn't be thousands, it should be hundreds. We can make larger versions as well, the size of buildings.'

However, he may have already been beaten to it. In 2018, the Shaanxi capital of Xi'an announced a very similar device towering up to 100 metres (330 feet) tall, around the height of a 30-storey building, called a Solar-Assisted Large-Scale Cleaning System (SALSCS). The idea was first proposed in a PhD thesis at the University of Minnesota in 2015. The Xi'an system consists of a large glass greenhouse with a 100m (330ft) chimney sticking out of it. The air inside the glass is heated up by the sun (just as it would be in a back garden greenhouse, only many times bigger) and as it rises, it is funnelled back out through the chimney fitted with a particle-trap filter, and clean air then comes out of the top. Short-term measurements show PM2.5 concentrations in the district surrounding the tower can be lowered by approximately 12 per cent. Models suggest that eight 500m-high (1,640ft) SALSCS towers (over 100m or 330ft taller than the Empire State Building) could reduce PM2.5 in Beijing by 15 per cent. A version covered with video advertising displays has been already been designed. It could be lifted from a *Blade Runner* film – smoky emissions continue side-by-side with giant filter towers, as glaring advertising films urge you to consume more and carry on.

When I attend the Smogathon London semi-final – a competition for start-ups with smog-busting business and technology ideas – in late 2017, I get the same uneasy feeling. The Smogathon London semi-final at the Google Campus is an intimate event, with maybe 30 people. The teams pace nervously in gleaming new trainers, waiting for their turn to pitch their ideas. Behind the stage a screen displays tweets from recent on-campus events, including one from Wikipedia founder Jimmy Wales, that reads, 'If you're never doing anything that fails, you just aren't innovating hard enough'. As the five-minute presentations get under way, a theme is quickly established: they are proposals to create small, paid-for pockets of clean air within urban environments. A pram with a £115 ($150) filter to fit around the baby; a car with an improved internal filtration system; a city bench that purifies the air for those sitting on it. They do nothing to tackle the underlying cause of air pollution, but do everything to mask it, to make it palatable. They strike me as the equivalent of the tobacco industry's first response to smoking health concerns: by adding a small white filter to the end, they said 'It's OK – you can keep on smoking now.'

Canned air, 30-storey smog towers and prams with inbuilt air filters are a sign of our folly, of how far down the wrong path we have already strayed. They also admit defeat. And admitting defeat remains a real option that some cities might settle for. In Delhi, Rana Dasgupta told me, 'a huge part of middle-class discussion about this has to do with masks and air purifiers … That you can buy your way out of the problem in various kinds of ways. It is totally absurd. Whereas being asked to drive less is seen as a huge affront. Most people I know basically just try to make sure that they have one car that is odd numbered and one car that is even numbered so they don't have to be in any way inconvenienced.' This, despite the fact that when we spoke the odd/even scheme had only been triggered twice, lasting for a total of 20 days. 'There is always a way that people just insulate themselves from the consequences,' said Rana.

Removing the sources of air pollution is not, however, the hardship option. It is an equally innovative, high-tech vision of the future, but it's one that doesn't put Mel Brooks's President Skroob in charge. Stanford's Professor Jacobson puts it most succinctly: 'If the world electrifies all sectors, and the electricity is obtained from wind, water and solar … then future progress will not result in any further air pollution.' (If you remember just one quote from this book, make it that one!)

Manufacturers now offer consumers cleaner options that are more attractive on their own terms than their rivals. Tesla's electric Model S displaced Mercedes to become America's best-selling luxury car in 2016, selling almost double its nearest rival by 2017. Bloomberg New Energy Finance predicts that the cost of ownership – including both the purchase price and running costs – of electric cars will dip below vehicles with internal combustion engines by 2022.

Volvo is aiming to only sell electric vehicles by 2025 – the same year that Norway, Austria and the Netherlands plan to achieve 100 per cent zero car emissions for new cars. Laura Parry from Islington Council, London, tells me that the zero emissions alternatives she offers to small businesses and residents are 'not just affordable, they will save you money. One guy I worked with … figured out that by using an electric van he is now saving £4,500 ($5,900) a year by not having to fill up with diesel … if you are spending less money on it and it's good for the air of your customers and your family … people are comfortable with that, they understand it.'

Consumer purchasing power can speed up the pace of change better than anything else. Antonia Gawel of the World Economic Forum writes, 'Each of us take decisions on a daily basis. We can drive or cycle to work. We can carry a reusable coffee mug or throw away a plastic equivalent every morning. We can ask questions about the food we buy or the clothes we purchase. While these actions may seem trivial at an individual level, at a collective level they will … grow the market for cleaner products and decrease demand for things that we know cause pollution.' The idea that zero emissions options are always more expensive is, frankly, nonsense. Most people can't afford a Tesla. But the end goal for any sane city is not for most people to own a Tesla (except maybe Oslo, where Tesla topped the new car sales charts in 2017). The end goal is to get more people walking, cycling, sharing public transport and joining car clubs – all far cheaper options than vehicle ownership.

David Newby, the cardiology professor who strapped cyclists into diesel exposure chambers, knows more than anyone the dangers of cycling through polluted air. Yet he diligently rides his own bicycle to work every day, up and down the steep Edinburgh hills, sharing the air with ageing vehicles belching blackened fumes. He insists that his children walk the mile to school and back rather than taking them in a car. He knows the air they are exposed to is unclean, he says. But it would be even more toxic were they to add another car to the streets. 'We need to find a way of getting people back into walking,' he says. 'If you go to a city where they have cleared out the cars and it's just people walking and cycling everywhere, my God it's such a nice place to live.'

So, the end goal is clear: 100 per cent renewable energy and zero emissions transport. But is the goal *zero pollution*? Even I would argue that's a step too far. For one, it's impossible – even large volumes of people walking and cycling will create some PM through resuspension of dust. The WHO sets a safe limit for PM of $20\mu g/m^3$, but the science increasingly shows there is no 'safe limit', only incrementally worse health effects for every 5 or $10\mu g/m^3$. I don't think you could ever get to the point of not making so much particulate matter that it wouldn't breach those [$20\mu g/m^3$] limits,' argues Ally Lewis. 'That's not a debate that people are currently having … I don't think yet we've got to the point of people saying well, OK, what level of pollution would be acceptable?'

Chris Griffiths at St Barts Hospital believes the answer is to 'go with the evidence of what is safe for people. People can have a happy life if they heat their houses in a different way – everybody likes an open fire, but if the health effects of that are demonstrably severe then we have to think about whether that's the right thing to do or not … When the health effects are known you then have to decide what to do about it. But if you're saying we need to live in a sort of Stalinist world where nobody is allowed to light a bonfire or have an open fire, that's not what we're after.' He draws the parallel with the ban on smoking cigarettes in bars and restaurants, which is now common in most countries. When we were kids, says Chris, being surrounded by cigarette smoke was just normal; now, nobody questions the ban: 'I don't think changing what is appropriate or allowed in public spaces is a problem, if the data is there to inform you.'

There are cultural red lines that some countries and cities may want to keep. The argument over fireworks during Diwali in India becomes heated because it is seen as part of Hindu tradition. As several Delhiites pointed out to me, firecrackers are a more modern tradition than many care to admit. But does anyone want zero fireworks on Diwali night? Or zero fireworks during Fourth of July celebrations in the US*, or 5 November in the UK? Do Australians want to ditch their barbecues in favour of electric hobs? It sounds flippant but it's an important point. I mentioned at the start of Chapter 4 that my surprise at the PM2.5 levels from frying my tortillas caused me to start baking them instead; well, in all honesty, I now fry them again, because they taste better cooked that way, but only when it's just me in the kitchen – while I can accept the personal risk-versus-taste gamble, I shouldn't expose my kids to it.

My favourite example came during my visit to Helsinki – a clean air city doing more than most to clean up what little pollution they do have. With one exception: saunas. Finns love their saunas, and most homes have at least one of them (there are an estimated 2 million saunas in a country of just 5.3 million people). A quarter of combustion-derived PM emissions in the Helsinki Metropolitan Area comes from wood burning, and half of that from wood-burning saunas. When I met Sampo Hietanen to talk about his transport ideas, I couldn't help sneaking in a question about this too: could Helsinki residents ever be persuaded to replace wood-fire saunas with electric ones? 'No. Because the wood sauna is just best,' he says. At first I wonder if he's joking. But then I realise he isn't. 'The Finnish fireplace manufacturers are quite innovative and have addressed this issue, like making new kinds of designs of the fireplaces to make them burn more efficiently … if it burns cleanly there is no problem.' There is, again, no such thing as clean burning. Burning solid fuel, by definition, creates emissions. But even Esa Nikunen, the head of the Helsinki Environment Centre which manages environmental protection across the city – and distributes leaflets promoting the benefits of electric saunas – told me,

---

* Home fireworks are already illegal in Los Angeles County, including for Fourth of July celebrations due to air quality and wild fire risks.

'the sauna is really a holy thing in Finland. Of course, it can be heated with electricity … but it's nicer if you burn wood.'

If a whole society is happy with that as a cultural red line, while being aware of the risks, then maybe that's OK? In general, however, societies (countries, cities, towns, villages) are rarely so homogenous. The most vulnerable and socially disenfranchised are disproportionately affected by air pollution. Often many of the polluters, especially individuals in their homes, simply don't know the damage they are doing due to a basic lack of awareness. Their choices need highlighting, debating and either re-affirming or scrapping, with equal input from all sections of society. But I refuse to believe there is any cultural or holy attachment to the refined versions of crude oil we pump into our vehicles' fuel tanks. Ultimately, if an alternative energy source can give us the same speed, journey time and comfort level that we've been used to, but for zero pollution from the exhaust pipe, then who cares what they run on? We know that electric and hydrogen cars can not only match petrol and diesel on all those points, but in fact beat them: electric motors can accelerate faster than fuel engines, they don't need gears, the ride is smoother and quieter. Better still, with the right cycling infrastructure, bicycles can be even quicker, healthier and far cheaper than electric cars. So, the final blueprint point should be made loud and clear: ban all petrol and diesel cars from cities as quickly as possible, and pay for their replacements by fining industrial emitters and the worst car industry cheats.

When I visited the British Museum, the final thing that Anna Davies-Barrett wanted to show me was a new display she had curated – her first, in fact – in the public collection of 1,000-year-old Sudanese skeletons with clear signs of the diseases they had suffered from, many linked to atmospheric pollution. Behind glass cases, bones and spines are warped, twisted and damaged in various ways, through diseases including TB and bone cancer: 'these things go hand-in-hand with asthma and chronic obstructive pulmonary disease,' says Anna, potentially there were other diseases they were suffering from too.' I ask her whether this helps to visualise the effects that present-day air quality is having on our own bodies, too? 'Yes, I'd like it to raise awareness about life now. I'd like people to be able to say, "look, they had all these respiratory diseases, that's awful, maybe we should change the way we think about air quality today?" ... If you are confronted by the physical evidence of something, if you can see the bone changes in the sinuses, bone changes in the rib, it is more shocking.' While writing this book, I was diagnosed with a condition myself (a very minor one, I should hastily add). Perennial rhinitis is, at least in my case, like mild hayfever, but year-round. After a particularly aggressive coughing fit one morning I thought I'd get it checked out, and there it was, written in black and white: 'occurs year-round and can result from ... air pollution such as automobile engine emissions'.

Air pollution is affecting our quality of life – everyone reading this book, everyone you know. Given the over-whelming links between air pollution and stunted childhood development, it is plausible – likely, even – that you and I would be healthier, more intelligent individuals than we are, had we not grown up breathing the pollutants that we have – the stubble-burning, the leaded petrol, the unregulated marine fuel. Remember the very first study at the start of this book, linking lead pollution exposure with the lowering of childhood IQ? Professor Bill Yule, who took part in UK lead pollution studies in the 1980s, recalled for a BBC Radio 4 documentary in 2018: 'Of course the argument that opponents ... would add is that "IQ" is very unreliable ... and it varies from morning to evening, and so a difference of four to five points [due to lead levels in the blood] is neither here nor there. I thought a lot about that ... If you imagine the normal distribution of IQ, the bell-shaped curve, and now imagine that it is shifted by four points to the left. In other words, lower. It makes little difference in the middle range. But in the lowest range, IQ 70 and below, there are at least double the number of children ... It is enormous.' The very last leaded petrol pump in the UK was removed in the year 2000. Prof Yule also said, 'There is a wide consensus now that there is no safe level. Back when I started working on lead almost 40 years ago, if you look in a paediatric textbook you would find that 60mg per decilitre [of lead in a child's blood] was the upper limit of normal … now population blood-lead levels have dropped to the point where we can ask the question, is 5mg per decilitre bad for children? Previously we were unable to ask that question because virtually all children had a level above 10mg per decilitre.'[7] As a child of the 1980s, I may have had blood lead levels above 10mg per decilitre, with the cognitive detriments that come with it.[*] Today's levels of diesel emissions, ultrafine particles and NOx could seem as astonishing to readers in the 2030s and 2040s as those leaded petrol figures seem now. And the children of today are exposed to higher levels of diesel emissions and nanoparticles than I ever was in the 1980s.

Children growing up in towns and cities could do so alongside roads of electric vehicles and cycle lanes, live in homes powered by renewable energy and breathe air almost entirely devoid of the pollution we take for granted today. This is an achievable vision. It is achievable right now in your city, in your town, in your back yard. Unlike climate change there is no '2 degrees' scenario, no knowledge that 'things are going to get worse whatever we do'. Urban air pollution is local, short-lived, and can be stopped at the source; the benefits are instant and dramatic. Whether this zero-emissions, low-carbon future happens in 10, 20 or 100 years, is down to public pressure and political will. It's down to us.

---

[*] I blame any mistakes you might find within this book on that, and that alone!

Epilogue

It is 21 June 2018. The longest day of the year and, possibly, personally my longest day of the year too. I've been up since five-something with the girls, both woken prematurely by the early sunrise, and am now on the long train journey to Manchester. It's the UK's second National Clean Air Day. My Egg is by my side for what is likely to be its last working tour of duty, before it gets put out to pasture. It is currently showing a $23\mu g/m^3$ reading for PM2.5 on my diesel train – higher than ideal, but not alarming – though it briefly got a heady rush of triple figures, like our days together in Delhi, as we changed trains in Birmingham New Street station.

The UK Environment Minister Michael Gove has just delivered an impassioned message over Twitter, wishing the UK a happy Clean Air Day, and asking us to think about how much we drive, how we heat our homes, how we take our children to school. People in Edinburgh, with the major thoroughfare The Mound closed to traffic for the day, are posting excited messages on social media and defiantly holding 'let's reclaim the streets' signs pushing for regular car-free days. The friends I made at Islington Council during last year's Clean Air Day in London are handing out 'free bikers' breakfasts' to a queue of happy cyclists – cycling in London is usually a thankless task, but this time they are being thanked. I share a message online from the UK Health Alliance, an advocacy group of healthcare professionals, which reads 'Did you know that by swapping 1 in 4 car journeys in urban areas, for walking or cycling, the UK could save over £1.1 billion in health damage costs!' I am on my way to experience Manchester's Clean Air Day and interview the mayor of Greater Manchester, Andy Burnham. My wife has just messaged me to buy more nappies. It is now 11.10 a.m. though, in truth, my day has barely begun. I can feel a headache developing. But this is my last trip for *Clearing the Air*, and I'm going to make the most of it.

I head first to the Manchester Royal Infirmary hospital, where a lung-testing tent has been set up for Clean Air Day. I arrive at midday and join a queue of nurses on their lunch break, keen for the free check-up. The test on offer is spirometry – blowing into a tube connected to a monitor no bigger than a phone, as hard as you can, for six seconds. A nurse in front of me is told that her lungs are smaller than average for her height and age, but that it is probably nothing to worry about. She looks concerned. 'Am I not normal, then?' she asks. The person doing the test reassures her that she is normal, that people have different lung sizes – but yes, hers are smaller than average. The nurse is in a rush to get back to work, her lunch break being almost over. She leaves, laughing to her friends, 'I'm not normal!' I ask if they have had many similar results and reactions today. A few, they say. One man had acute shortness of breath, so they recommended he see his GP. When it's my turn, I take a deep breath and blow. Six seconds is a long time, it feels like all your lungs can manage is three seconds, but I am told to keep going. I feel dizzy. But I'm told my lung volume is good, in the 88th percentile. I am given a score for my forced expiratory volume (FEV1), and forced vital capacity (FVC), both terms I remember from the California children's lungs studies in Chapter 6. I think of all the children with lungs permanently reduced below normal capacity, simply because of how close to a road they live. I wonder if that nurse is one of them. Along with my test results I receive a Clean Air Day wristband as a freebie, and, bizarrely, a car air freshener. 'It's maybe not the most suitable idea for Clean Air Day!' laughs one of the volunteers. I find a nearby bench to sit down and write, and realise I am sitting next to a heavily pregnant woman and her partner. The labour ward must be nearby. I ask them if it their first child. It is. I wish them the best of luck and realise that I am back where my journey began, four years ago when my parenting and clean air journey began.

I decide to walk the 30 minutes it will take me to get from the hospital to Exchange Square, where I'm due to meet the mayor. The walk takes me along Oxford Road, a recently revamped major thoroughfare with added segregated cycle lanes and a ban on private cars in daytime hours. I only see buses and cyclists. My Egg for a time reads $1\mu g/m^3$, something I haven't seen in the centre of a major city since Helsinki. It lingers around single figures most times I glance at it during my walk. Signs advertising the university, reading 'How Will You Change The World?', hang high on lamp posts like celebratory flags. Despite the healthy flow of buses, cyclists don't have to stop or slow at the regular bus stops, because the cycle lanes have been designed to swerve behind bus stops. The bus drivers travel unimpeded by cars on the road, but it is the cyclists who have the speed advantage – everyone is benefitting.

In Exchange Square, there is another lung-testing tent, various electric cars and bicycles to try, volunteers handing out Clean Air Day leaflets, and in the middle of it the mayor surrounded by a scrum of TV crews. I get a flashback to Paris, and unexpectedly finding myself in the middle of Anne Hidalgo's and arrival in the Place de la Bastille – but this time I actually have an interview slot. Andy has himself recently returned from a meeting with Hidalgo, in Paris, where he signed a pledge to move his city towards a zero emission bus fleet. He admits to being inspired by her work, and that of Sadiq Kahn in London. But he hopes to supersede both their efforts and make Greater Manchester – a conurbation of some 2.7 million residents and 500 square miles – a global leader in air quality and low carbon innovation. His ambition is to make Manchester 'carbon neutral 10 years ahead of the rest of the country,' he tells me, when we sit down together. 'That means lots of change, not just in transport, but buildings, energy. We have to give a different proposition to the public, whereby they think about buildings and cars in terms of lifetime cost as opposed to purchase cost. If you have a zero-carbon house, you might baulk at the price of it, but your heating and fuels bills might be a fraction of what you're used to. Once you've got your electric car, it is cheaper to run … we are thinking about setting a date in our plan whereby all new homes have to be zero carbon. The point about that is, if you do something like that it moves the market.' I tell him about the French law, in place since 2015, that requires new commercial buildings to have either solar panels or green roofs. 'That is effectively what we are looking at too,' he confirms. 'We need to be zero carbon … because we think there will be economic benefits that come from being the place that moves first. There are two big drivers in the twenty-first-century economy: digitalisation and decarbonisation. The cities that are leaders in both will be the most prosperous.'

Some of that, of course, is polished rhetoric from a seasoned politician. What we need now more than ever is action, not words. I present him with a printout of my Clean Air Blueprint for Cities (p. 305), and tell him about my frustration with the speed of change. There are no big surprises in this blueprint, I say, we know what needs to be done. Why can't we – why can't *he* – get on and do them already? 'The solutions are there,' he agrees, as he scans the blueprint. 'It's not that complicated. The will is there. Give us the tools, we will do it. Really. I would tomorrow … This is an issue of health inequality and health injustice. That's really important. On what basis is it fine for people with massive BMWs to be polluting the air of kids in poor areas? That's not fine. And it's not fine for cowboy bus companies to be running old diesel buses past the school gates.' And yet in London 37 per cent of all buses are electric or meet Euro 6 standards, whereas, as we speak in mid-2018, Greater Manchester has just 10 per cent – that surely is something he can address, and quickly? 'The debate is when not if – we haven't yet got the financial mechanism to make it happen at speed. In the 1980s we were told that deregulation [when bus transport was moved from a state-owned service in the UK to contracts for private operators] would lead to more services and lower prices. The total opposite happened. My daughter yesterday spent £4 to travel four miles by bus. What's the bus fare in London [where transport remains state-owned, by TfL]? £1.50. Deregulated buses have been a race to the bottom – they pack out the lucrative routes where they create illegal air because they are all nose to tail. And the state of the railways is so bad in northern England that people are going back to using their cars.' Since May 2017, however, Greater Manchester has acquired newly devolved powers from the UK government, including tighter control over transport. 'We've got the power now to regulate the buses. And I've already told them that moving to zero emissions buses will be the next step.' I ask him if he'd heard of the electric bus in Milton Keynes that wirelessly charges en route, that doesn't need to stop to recharge, and has been up and running for three years. He looks surprised. No, he hasn't. And not for the first time, I wonder at the lack of communication between cities even within the same country, let alone internationally.

Burnham is, quite obviously, a left-wing politician – he was a former minister in the Labour government, and MP for his home town, Leigh, a former mining community within Greater Manchester. So I ask him something that has been bothering me – and bothered Devra Davis's husband Richard, in Chapter 10 – about the cost of clean air action. As convinced as I am about the benefits to society being a net gain, there is still a question over where that initial money comes from, and a question over how you ask the poorer members of society to switch their car to electric, or upgrade their housing and heating to renewable sources. 'The area where I live is a [former] mining community,' he says. 'But in some ways the understanding there is greater – those industries left a legacy of respiratory disease. I can remember as an MP, it was a very common experience to meet constituents, the older miners, strapped to two oxygen bottles. Because of that there is an understanding of the damage to people's lungs from breathing in dust and particulates. Maybe it's the poorest areas that have a better memory of poor air and what it can do.' To pay for it, he wants a national vehicle scrappage scheme to replace older polluting vehicles and that supports drivers and businesses to change to low-emission vehicles and other sustainable forms of transport, weighted towards those on low incomes. This 'Clean Air Fund' would be jointly funded by government and car companies. But most importantly for the people in his city, he says, moving towards electrified transport and low carbon homes will 'save people money. That's when this agenda will really move – when it's understood that green living is cheaper living.'

After speaking with Mayor Burnham, I check out the other Clean Air Day stands in the city centre. I haven't been on an electric bike since the last Clean Air Day in London, and am not desperate to do so again, but Pavol from Manchester Bike Hire talks me into it. Between the Lung Health Dome and promotion stands for electric cars, Pavol's small company is trying to persuade businesses to use his bikes as delivery vehicles instead of vans. His team of couriers ride electric-assisted cargo bikes all over Manchester, delivering for major companies and local businesses. He says the bikes are cheaper than the alternatives, quicker (his bikes can access areas that vans cannot), cleaner ... but although he has a healthy business, he is frustrated by the numbers of meetings he's had with big companies who nod and agree with everything he says but, when it comes to the crunch of re-contracting, stick with what they know – a man in a diesel van. The fleet of cargo bikes Pavol shows me are not shiny and new, but rugged and used, and the better for it – they scream robust practicality. One particularly lumpy-looking one with a solid metal goods box strapped to the front is capable of loading and moving 250kg (550lb), a quarter of a metric tonne, in one go. Pavol insists that I take it for a spin on the busy square – I try to resist, worried that this thing will topple over, crushing me and some unfortunate bystander, but he insists he's recommending it *because* it is easy to handle. So I do. And he's right. It is a clean, green (actually, blue) beast of burden, feeling safer than an ordinary bike in the same way that driving a Land Rover feels safer than a hatchback. And, like my previous experience of riding an electric bike in Greenwich, it makes me feel like Superman when gliding uphill. I now very much want to buy one to load my children into for the school run.[*]

---

[*] Probably without telling them about the electric-assisted part as I breeze effortlessly uphill.

My visit to Manchester also gives me my first in-car Tesla experience. I've written about them for years, but never actually sat in one. The two models on display, the X and the S, feel exactly as I imagined they would – like a Transformer that has turned from a smartphone into a car. It is a lush, digital experience. Huge windscreens curve over and above the driver's seat, giving a sense of bringing the outside in – not the usual insular experience of many sports cars. I wonder if this is in preparation for that eventual, inevitable, fully autonomous upgrade, when the steering wheel becomes obsolete and the driver is free to stare up at the clouds. There is a BMW i and the latest Nissan Leaf on display in Exchange Square too, but I just don't have the same urge to give them a spin. They simply look like cars. Tesla is the superstar, the one the teenagers are queuing to sit in and take selfies with. And even though some of the sheen has since gone from Elon Musk's appeal – the sideline in flamethrowers and mass worker lay-offs come to mind – I still applaud what Tesla has achieved by making electric a bling, must-have. I love its star-status sex appeal. I say that even someone who will probably never be able to afford one, but simply acknowledge that its appeal will make me, and others, buy* the nearest thing they can afford. And that will probably be a second-hand Nissan Leaf (and I applaud Nissan just as much for ploughing on with a mass-produced, unsexy but affordable electric car, before there was even a market for one).

Manchester reminds me a lot of Helsinki. The same legacy of electric trams is now paying off big time. It shows in the readings on my Egg (I take a taxi at one point and put my Egg on the seat beside me, the first time I remember doing so since Delhi. The reading is $3\mu g/m^3$); and the geographical advantage, too, of being in a cool, windy and rainy part of the world. But that isn't the full story of a clean air city. The story is one of urgency and desire. Like Helsinki, Manchester also plans to turn its citizens into cyclists. Chris Boardman, a former Olympic cycling champion and Greater Manchester's

---

* Or more likely share, rent or lease.

Cycling and Walking Commissioner, has proposed a city-wide cycling and walking network made up of more than 1,000 miles of routes, including 75 miles of Dutch-style segregated bike lanes. It also includes 1,400 safer road crossings and previous concrete wildernesses revamped as green public spaces to sit and play.

The city that made the biggest impression on me, however, will always be Beijing. An air pollution pariah, in international terms. The number one most-polluted city in the world. The capital of a mega-power building its riches on a foundation of industry, irrespective of the pollution cost. And then it pulled the handbrake, stopped dead, and said 'we can do this another way'. If Beijing, the scene of the 2013 Airpocalypse, can clear the air, then anywhere can. There are no excuses any more. No city has a unique get-out-of-jail-card excuse for why its air pollution is somehow different and unavoidable. The history of air pollution is one of constant, circular repetition. Fortunately, that includes many repeatable lessons for how to improve. As Andy Burnham told me, 'Manchester led an industrial revolution two centuries ago that took away people's clean air. Now we've got to lead one to give it back. That's the way to put it across to people: we can build future prosperity at the same time as cleaning everything up.' I leave Manchester that evening – on a packed diesel commuter train, carrying a car air freshener, a bag of nappies and book deadline nerves – hopeful for the future. Impatient, but hopeful.

The Clean Air Blueprint: For Cities

- Ban all petrol and diesel cars from city centres as quickly as possible – copy the Paris Crit'Air scheme as a means of doing so – and pay for it, and the points below, by fining the biggest emitters.
- Replace fleets of diesel buses with electric buses.
- Invest in walking and cycling infrastructure, specifically pavements, protected cycle lanes (physically separated from traffic) and bicycle parking.
- Pedestrianise your major shopping streets during daylight hours – research shows that local businesses benefit as footfall is increased.
- Green your city: plant trees, protect parks and require buildings to install green roofs.
- Install green walls besides busy roads, schools and hospitals.
- Set a renewable energy target. Aim for 100 per cent renewable energy procurement by 2030, with a step-by-step five-year plan for how to get there. Require planning permission for new buildings to include on-site renewable energy such as solar panels.
- Greatly increase the number of electric car-charging points. Consider the retro-fit of existing infrastructure such as lampposts on every residential street.
- Make city parking free for electric cars, in the short term.
- Ban solid-fuel burners such as wood-burning stoves and coal fires for domestic heating within built-up areas (unless alternative affordable heat sources are unavailable – in which case, sort that out first).

- Work with train authorities and national government on plans to move towards full railway electrification, replacing the diesel fleet.
- Port cities: require all ports to install 'cold ironing', which plugs ships in to local power supplies while in port, rather than running their engines.
- No matter how good your clean air laws, back them up with enforcement; properly fund enforcement authorities.
- Implement 'environmental justice' schemes which identify the communities that suffer from the worst pollution, and weight the investment of all the above points in favour of those communities.

The Clean Air Blueprint: For You

- Read the blueprint for cities and start lobbying your local politicians – how are they doing on all those points? And if they are not doing any of them, why not? If they say they don't have the budget, ask if they fine polluters and how that money is spent.
- Check out the government's air quality measurements near you. Get an app that gives real-time air quality readings. Consider getting your own portable pollution monitor. Either way, educate yourself about the real pollution levels where you live and work.
- Before you burn anything – whether it's fuel in a car or charcoal on a fire – ask yourself if you really need to, or if there is a viable zero-emissions alternative.
- Try to walk, cycle or use public transport rather than using cars for short journeys (and if you're unfit or lazy, like me, consider an electric bicycle or scooter).
- If you live in a city, join an electric car club. If there isn't one, write to one and tell them to come to your city.
- If you need to buy a car, check out the electric cars on the market (especially the second-hand electric and hybrid models) and the government grants available.
- Switch your home energy supplier to one that offers 100 per cent renewable electricity.
- Consider making your own renewable heat or energy, such as solar panels or ground or air-sourced heating.
- Do not – I repeat, do not – install a wood-burning stove, if you live in a built-up area, no matter how

'eco' the marketing brochure claims it is. If you live in a log cabin in a forest, then fine, burn away.

- Talk to your local schools and nurseries about installing green walls and implementing 'walking bus' schemes (see p. 245).
- If you live beside a busy road, or a pollution source such as a diesel railway or diesel generators, create your own green wall or green roof. There's lots of advice online about how to do it, but a traditional privet or conifer hedge does a good job.
- Don't panic! Even if you live in a horribly polluted city, you can reduce your personal exposure. Walk or cycle on back streets rather than main roads. If you have to walk along a major road, walk on the building side of the pavement rather than the roadside kerb to reduce your nanoparticle exposure.
- Masks, air filters, pram filters – honestly, it's your call. They can help. If nothing else, they raise awareness and get people talking about the issue. But really, unless you live in a filtered environment all day long, you're going to be exposed to pollution. So, do all the points above first. But by all means, do everything you can to protect yourself.
- Tell others about this book (shameless, I know!)

References

Many more studies are of course referred to in this book, but for reasons of space I can only give full references for those I quote from in most detail.

Prologue

1 Wright. J. P., *et al.* 2008. 'Association of Prenatal and Childhood Blood Lead Concentrations with Criminal Arrests in Early Adulthood', *PLoS Medicine*.
2 'Our Lives with Lead', *In Their Element*, BBC Radio 4, January 2018.

Chapter 1: The Greatest Smog?

1 Berridge, V. and Taylor, S. 2005. *The Big Smoke: Fifty Years after the 1952 London Smog,* Centre for History in Public Health, London School of Hygiene & Tropical Medicine.
2 Wallis, H.F., *The New Battle of Britain*, Charles Knight & Co, 1972.
3 'Report of the Special Rapporteur on the implications for human rights of the environmentally sound management and disposal of hazardous substances and wastes on his mission to the United Kingdom of Great Britain and Northern Ireland', UN Human Rights Council, September 2017.
4 'The NDTV Dialogues: Tackling India's Killer Air', NDTV (New Delhi Television Ltd), 19 November 2017.
5 'Heart attacks, respiratory diseases, cancer, NCDs cause of six out of 10 deaths in India', *Hindustan Times*, 14 November 2017.
6 'Pollution stops play at Delhi Test match as bowlers struggle to breathe', *Guardian*, 3 December 2017.
7 Davis, D. 2002. *When Smoke Ran Like Water*, Basic Books.
8 L'Enquête Globale Transport, Observatoire de la mobilité en Île-de-France (OMNIL), 2013.

Chapter 2: Life's A Gas

1   Baumbach, G., *et al.* 1995. 'Air pollution in a large tropical city with a high traffic density – results of measurements in Lagos, Nigeria', *Science of the Total Environment* 169.

2   Lewis, A. 'Air Quality and Health', University of York lecture [slides as seen in 2017].

3   Fioletov, V. E., *et al.* 2016. 'A global catalogue of large sulphur dioxide sources and emissions derived from the Ozone Monitoring Instrument', *Atmospheric Chemistry and Physics*.

4   Sahay, S. and Ghosh, G. 2013. 'Monitoring variation in greenhouse gases concentration in Urban Environment of Delhi', *Environmental Monitoring and Assessment*.

5   Sindhwani, R. and Goyal, P. 2014. 'Assessment of traffic-generated gaseous and particulate matter emissions and trends over Delhi (2000–2010)', *Atmospheric Pollution Research*.

6   Landrigan, P. J., *et al.* 2018. 'The *Lancet* Commission on pollution and health', *The Lancet*.

7   Davies, S. 2018. *Chief Medical Officer annual report 2017: health impacts of all pollution – what do we know?*, Department of Health and Social Care.

Chapter 3: Particulate Matters

1   Wang, Z., *et al.* 2013. 'Radiative forcing and climate response due to the presence of black carbon in cloud droplets', *Journal of Geophysical Research: Atmospheres*.

2   Davis, D. 2002. *When Smoke Ran Like Water*, Basic Books.

Chapter 4: No Smoke Without Fire

1   Lewis, M. E., *et al.* 1995. 'Comparative study of the prevalence of maxillary sinusitis in later Medieval urban and rural populations in Northern England', *American Journal of Physical Anthropology*.

2   Seinfeld, J. H. and Pandis, S. N. 2006. 'Atmospheric Chemistry and Physics: From Air Pollution to Climate Change', Wiley.

Chapter 5: The Dash For Diesel

1   Vidal, J. 'All choked up: did Britain's dirty air make me dangerously ill?', *Guardian*, 20 June 2015.

2   'Fine Particles and Health', POST Technical Report, June 1996.

3   Laxen, K. 'Will backup generators be the next "Dieselgate" for the UK?', *environmental SCIENTIST*, April 2017

4   Kumar, *et al*. 2011. 'Preliminary estimates of nanoparticle number emissions from road vehicles in megacity Delhi and associated health impacts', *Environ Sci Technol*.

5   Li, N., *et al*. 2016. 'A work group report on ultrafine particles (American Academy of Allergy, Asthma & Immunology): Why ambient ultrafine and engineered nanoparticles should receive special attention for possible adverse health outcomes in human subjects', *Journal of Allergy and Clinical Immunology*.

Chapter 6: Struggling To Breathe

1   Gauderman, W. J., *et al*. 2015. 'Association of Improved Air Quality with Lung Development in Children', *New England Journal of Medicine*.

2   Findlay, F., *et al*. 2017. 'Carbon Nanoparticles Inhibit the Antimicrobial Activities of the Human Cathelicidin LL-37 through Structural Alteration', *Journal of Immunology*.

Chapter 7: The Greatest Smog Solution?

1   Nguyen, N. P. and Marshall, J. D. 2018. 'Impact, efficiency, inequality, and injustice of urban air pollution: variability by emission location', *Environmental Research Letters*.

Chapter 9: Road Rage

1   Tainio, M., *et al*. 2016. 'Can air pollution negate the health benefits of cycling and walking?', *Preventive Medicine*.

Chapter 10: What Price, Fresh Air?

1   Trasande, L., *et al.* 2016. 'Particulate Matter Exposure and
    Preterm Birth: Estimates of U.S. Attributable Burden and
    Economic Costs', *Environmental Health Perspectives*.

2   Chang, T., *et al.* 2016. 'The Effect of Pollution on Worker
    Productivity: Evidence from Call Center Workers in
    China', *National Bureau of Economic Research*.

3   Zivin, J. G. and Neidell, M. 2012. 'The Impact of Pollution
    on Worker Productivity', *American Economic Review*.

4   Singh. S. 'Metro Matters: Delhi can't grudge what it takes to
    breathe easy', *Hindustan Times*, 23 October 2017

5   Jacobson, M. Z., *et al.* 2015. 'Low-cost solution to the grid
    reliability problem with 100% penetration of intermittent
    wind, water, and solar for all purposes', *PNAS*.

6   Jacobson, M. Z., *et al.* 2017. '100% Clean and Renewable
    Wind, Water, and Sunlight All-Sector Energy Roadmaps for
    139 Countries of the World', *Joule*.

7   'Our Lives with Lead', *In Their Element*, BBC Radio 4,
    January 2018.

Acknowledgements

First, foremost and forever, I thank my wife Dr Patricia Brekke. Without her support, wisdom, patience, and occasional basic science tutelage, this book wouldn't have happened. From the whole of my heart, I thank you.

Thanks to my agent, Jenny Hewson at Rogers, Coleridge & White, for taking me on and giving me the self-belief and kick-up-the-arse I needed to get my book proposal into shape.

For helping this book grow from an idea into a distinct possibility with their time, input and advice, I would like to thank the superb science author Alanna Collen, Tim Reid at ClientEarth, and Carrie Plitt.

For taking a chance on a first-time author, and for their thoughtful, insightful editing, I owe huge thanks to Jim Martin and Anna MacDiarmid at Bloomsbury.

During the writing and research, I have too many people to thank to list them all – if you are quoted in the book, I hope you will accept that as acknowledgement enough (let's face it, readers are more likely to see your name there than they are here!). But some special mentions: my brother-in-law Ewan Jones, for sending me newsletters, links, hastily taken photos of information posters, and his impeccably patient (and detailed!) maths and engineering explanations. To John Saffell at Alphasense for finding out what I "was up to" and giving me some valuable early steers. To my near-neighbour Neil Wallis, who made some very useful introductions, not least to his father's brilliant book *The New Battle of Britain*. To transport guru Lukas Neckermann for infecting me with the electric and autonomous travel bug. And Suzanne Williams for the transcription services and (equally valuable) sense of humour.

To my hosts on my travels: in Delhi, Vandana and her father Vinod, for their kind hospitality and generous help during my stay. In Beijing, I cannot thank Manny Menendez enough for his time, guidance, and hugely entertaining company. In Paris, thanks to my good friend Annabelle Sappa (and cat) for the lodgings and *bonhomie*.

For help and support as the book deadline was getting agonisingly near yet so far, thanks to Hannah Gattrell. And, of course, to all my friends and family, who have had to put up with me droning on about this subject for the last two years – and, now it's published, for a little while longer, too...

Finally, although I have already dedicated this book to my children and nephews (it's obvious, really – you are the future), I want to add another dedication at the end to two friends who passed away, much too young, whilst I was writing this book in the spring of 2018: Richard Silver and Ben Collen. You both continue to inspire.

Index